# PROBLEM-SOLVING KIDS

Creating self-directed, problem-solving children

By Barbara F. Backer, M.Ed. and Susan A. Miller, Ed.D.
Illustrated by Gary Mohrman

Totline® Publications
A Division of Frank Schaffer Publications, Inc.
Torrance, California

**Managing Editor:** Kathleen Cubley
**Editors:** Elizabeth McKinnon and Gayle Bittinger
**Contributing Editors:** Carol Gnojewski, Susan Hodges, Susan Sexton, Jean Warren
**Copyeditor:** Kris Fulsaas
**Proofreader:** Miriam Bulmer
**Editorial Assistant:** Durby Peterson
**Typist for Susan Miller:** Karen Epting
**Graphic Designers:** Sarah Ness and Gordon Frazier
**Graphic Designer (Cover):** Brenda Mann Harrison
**Production Manager:** Melody Olney



ISBN: 1-57029-288-4

Printed in the United States of America
Published by Totline® Publications.
23740 Hawthorne Blvd.
Torrance, CA 90505

# INTRODUCTION

Throughout the day, problem-solving opportunities abound for young children (and all of us!). These opportunities challenge your children's critical thinking skills as they analyze a situation and decide what to do. They also encourage creativity as children often combine ideas in new ways or produce original concepts.

In *Problem-Solving Kids*, over one hundred problem-solving scenarios involving six different areas and environments—art, dramatic play, manipulatives, blocks, science, and outdoors—are featured. The meaningful problems come from children's real play and reflect their points of view. The scenarios provide a springboard from which children can actively discover how or why things work, rather than having adults provide the solution or explanation. The activities help them build skills like creative thinking, critical thinking, motor, and social-emotional.

Various problem-solving steps are suggested to help your children identify problems and then brainstorm workable solutions. Your role in encouraging this process is outlined as well. There are no right or wrong answers—all of your children's ideas should be respected. When they try something and it doesn't work, this, too, is a valuable learning experience. It allows them the chance to analyze and evaluate another possible solution, and then try again.

This book is filled with a wonderful variety of open-ended opportunities that will foster curiosity; encourage creativity; promote critical thinking, social interactions, and dialogue; and strengthen physical skills. Together, these skills and practices lay an important foundation for your children's lifelong problem-solving skills.

# CONTENTS

# STEPS FOR PROBLEM SOLVING

*Below are steps that you can take with your children when determining problems, developing solutions, and putting those solutions into action.*

1. **IDENTIFY** – Determine and discuss the problem. It should be meaningful, interesting, and appropriate for young children.

2. **BRAINSTORM** – Encourage your children to think about possible solutions. Listen to and respect all of their ideas. Keep a record of the solutions suggested in case the children want to try more than one.

3. **SELECT** – Help your children examine the advantages and disadvantages of various solutions and then choose one that seems workable.

4. **EXPLORE AND IMPLEMENT** – Let your children gather the necessary materials and resources and then try out the solution they decided upon.

5. **EVALUATE** – With your children, observe and discuss whether the solution to the problem was successful. If appropriate, help the children think of changes they might want to make in the idea they tried. Or, encourage them to try other solutions.

# YOUR ROLE

*Your role in facilitating the development of your children's problem solving behaviors is that of encourager and guide. Be sure to allow plenty of time and space for the children to brainstorm and try out their ideas. Below are other ways you can help.*

- Observe carefully to see how you can facilitate the problem-solving process. For instance, provide or add special materials when needed. Also, encourage your children to observe in order to verify what is happening.
- Turn any mistake into a learning experience. Help your children analyze why a solution did not work out.
- Promote discussion, interaction, and collaboration. As your children work together, help them to be patient and persistent in their explorations.
- Throughout the day, model problem-solving skills in natural ways. However, avoid solving your children's problems for them. Trust them to come up with solutions!
- Listen to, acknowledge, and support your children's ideas. Respect their suggestions and encourage them to value one another's viewpoints.
- Provide easy access to stimulating materials. If your children need assistance, help them learn how to use the materials in diverse ways.
- Arouse curiosity. Ask open-ended questions that begin with "What if . . . ?" or "Why do you suppose . . . ?" to foster creative and critical thinking.
- Encourage experimentation. Let your children know that it is all right to take a guess or try a new idea (always keeping in mind the children's safety). Help them understand that there are no "perfect" solutions.

# Problem Solving in the
# ART AREA

# UNEXPECTED RUBBINGS

*A boy is coloring at the art table. As he works, an outline, or rubbing, of a shape appears on his paper. Your children want to know how this has happened so that they can make rubbings too. What can they do?*

| Children's Possible Solutions | Your Role |
|---|---|
| **1.** The boy decides to color all over his paper to see if more shapes appear. | Provide crayons in a variety of colors. Encourage the boy to continue exploring. |
| **2.** A girl suggests that the boy try coloring with different crayons. | Encourage the child's actions. Ask him to describe what is happening. |
| **3.** Other children say, "Look under your paper." When the boy does so, he discovers a paper shape that matches the rubbing. Your children find other paper shapes in the scrap box and decide to use them to try making their own rubbings. | Make sure that the scrap box contains a variety of paper shapes as well as shapes cut from different-textured paper. |

## Extending the Experience

- Make rubbings of a variety of items, such as coins, ribbons, doilies, or textured placemats.
- Outdoors, make rubbings of items such as tree bark, leaves, cement walkways, trike pedals, or swing chains.
- Make rubbings of the bottoms of the children's athletic shoes.

# HOLES IN THE PAPER

*Several children are reclining on the carpet, drawing pictures on paper. Their pencils are pushing through the paper and making holes. How can the children continue with their artwork?*

| Children's Possible Solutions | Your Role |
|---|---|
| 1. One child decides to repair the holes in his paper with tape. | Have tape available for your children's use. Help them discover that placing tape on the back of their papers hides the repairs and keeps the tape from covering their pictures. |
| 2. Another child wants to put a book or a magazine under her paper to provide a firm surface. | Support this suggestion by having magazines and other hard, portable surfaces available. The covers of discarded wallpaper sample books work well. |
| 3. Other children decide to incorporate the holes in their papers into their artwork. The children try using other items, such as crayons and chalk, to see if they will make holes too. | Support your children's efforts and have safe items available. Ask, "Which items make the biggest holes? The smallest? Which work best? Why?" |

## Extending the Experience

- Look around for different kinds of holes. Some examples are holes in lace-up shoes, window screens, fences, suitcase handles, hoops, and rings.
- Make sieves by punching holes in various materials, such as pieces of aluminum foil, sheets of waxed paper, or cardboard shoeboxes. Which materials are easiest to punch holes in? Which work best as sieves?
- In the sides of large cardboard cartons, cut holes for your children to crawl through.

# MASHED-IN MARKER TIPS

*Your children take out felt tip markers to use for their artwork. But when they try to draw, they discover that the markers won't work because the tips are all mashed in. What can the children do?*

| Children's Possible Solutions | Your Role |
|---|---|
| 1. In an attempt to pull out the marker tips, several children dig at them with their fingers. | Encourage your children's efforts and praise their persistence. |
| 2. Other children suggest trying to dig out the tips with scissors. | Acknowledge the children's solution. Then discuss scissor safety and how scissors should properly be used. |
| 3. One child says, "We need new markers." | Facilitate discussion about how the markers might have become damaged. How will the children protect new markers? |

## Extending the Experience

- Use mashed-in markers to make indentations when playing with modeling dough.
- Mash golf tees into modeling dough that has been placed in a large margarine tub.
- Recite and act out finger plays that include hammering motions.

# SMEARED ART PASTELS

*When your children use art pastels, the colors often smudge and rub off on their hands and clothes. How can they prevent this?*

| Children's Possible Solutions | Your Role |
|---|---|
| **1.** One child decides to paint a layer of glue over his creation to make the colors stay put. | Encourage the child's problem solving by providing brushes and glue. Help the child evaluate his method while the glue is wet and again when the glue is dry. |
| **2.** A girl decides to paint over her drawing with water to make colors that will not rub off. | Encourage the child by providing water and brushes. Help her evaluate her method while the paper is wet and again when it is dry. |
| **3.** A boy tries to wipe off the excess dust by rubbing his drawing with a wad of facial tissue. This creates a smeared abstract picture. | Encourage the boy's efforts. Ask him to describe what happened to his drawing and help him evaluate his method. |

## Extending the Experience

- Use sidewalk chalk to draw pictures outdoors. Observe over time to see if the colors remain sharp and bright.
- Dip art pastels in water, then use them to draw on dark-colored construction paper. Compare the results with drawings made with dry pastels.
- Use art pastels to draw on wet construction paper. Does the color smudge when the paper is wet? When the paper dries?

# PAPER STRIPS GALORE

*Your children find a lot of paper strips left over from another art project. What can they do with the strips?*

| Children's Possible Solutions | Your Role |
|---|---|
| **1.** Several children suggest using the strips to make a paper chain. | Encourage the children's efforts. Offer to display the chain for everyone to admire. |
| **2.** Another child demonstrates how to make 3-D art by gluing the ends of several strips to a cardboard base and then twisting the strips before gluing down the other end. He weaves additional strips over and under the glued strips. | Help the child describe how he constructed his creation. |
| **3.** A girl demonstrates how to staple several strips together at angles to make a long zigzag. She uses this on the floor as a road for toy cars. | Compliment the child on her creative idea. |

## Extending the Experience

- Make a paper chain calendar. Count the links to see how many days until a special event, such as a field trip. Each day, remove a link of the chain. On the day of the big event, remove the final link.
- Using strips of colored paper, make pattern chains, such as yellow-green-yellow-green or red-red-blue-red-red-blue.

# A STAINED WALL

*A wall in your room has stains on it. Your children all agree that the stains are unsightly. What can they do?*

| Children's Possible Solutions | Your Role |
|---|---|
| **1.** Several children think that they should try washing the stains off the wall. | Provide buckets of water and sponges along with towels for mopping up any spills. |
| **2.** Other children suggest making a large mural on paper to hang over the stain. | Facilitate discussion about the size, color, and topic of the mural. Provide materials and encouragement. When the mural is completed and hung, invite parents and other classes to come see how different the wall looks. |

## Extending the Experience

- Use large house-painting brushes and buckets of water to "paint" the outside walls of your center.
- Make other murals. Use them to decorate inside doors, walls outside the room, and bulletin boards.

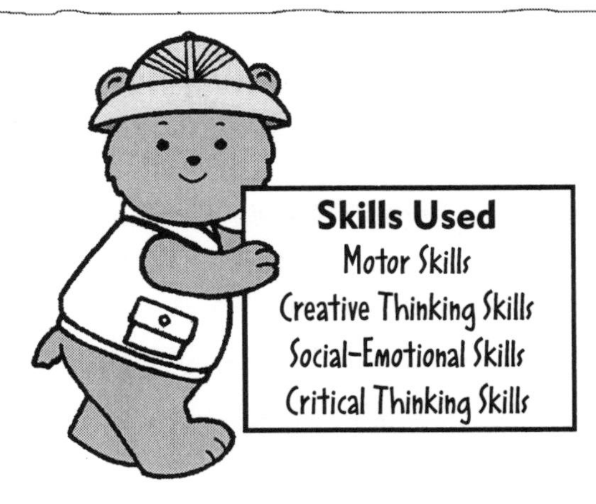

**Skills Used**
Motor Skills
Creative Thinking Skills
Social-Emotional Skills
Critical Thinking Skills

# WHAT A MESS!

*The art table is extremely messy with marker streaks, wet and dried glue, small pieces of modeling dough, and other materials stuck to it. What can your children do?*

| Children's Possible Solutions | Your Role |
|---|---|
| **1.** Several children try using sponges and brushes from the water table and hand soap from the bathroom to clean the art table. | Encourage your children's efforts. Afterward ask, "Did the cleaning materials work? What else might we try?" Provide safe materials, and caution the children to clean up spills. |
| **2.** One child suggests making a rule: "People who mess up must clean up!" | Have your children discuss this idea. How will they enforce the rule? Provide any materials the children request to facilitate their idea, such as a sign-in sheet or a poster stating the rule. |
| **3.** Other children suggest covering the art table with newspaper early each day. They think this will protect the table and make cleanup easier. | Remind your children to bring in newspapers, and provide a place to store them. Help the children decide on a fair way to delegate the daily job of covering the table. |

## Extending the Experience

- Provide materials to encourage cleaning all parts of your room. Include items such as a small broom and a dustpan, a small mop, a feather duster, a non-scratch scouring pad, and nontoxic cleaners.
- Encourage outdoor cleanup too. Pick up trash in the play yard and sweep walkways.

# A BIG, BIG BOX

*A parent has brought in a large appliance box for your children to play with. How can they use it?*

| Children's Possible Solutions | Your Role |
|---|---|
| 1. Your children suggest ways to use the box: as a fort outdoors; as a hiding place outdoors; as a quiet reading place indoors; as a small playhouse. | Discuss all ideas. Ask, "Since there is only one box, how can we decide which idea to choose?" |
| 2. Several children suggest finding more boxes so that all of the desired items can be made. | Bring in more boxes of all sizes. (Furniture and appliance stores and office supply companies will donate large cartons.) Discuss with your children how they will use the boxes, and provide materials such as felt tip markers, tape, and glue. |
| 3. Several other children want to use the original appliance box plus any additional cartons to make a long, long tunnel. | Brainstorm ideas for accomplishing this idea. How can the children make the tunnel long? Offer assistance if they need help. |

## Extending the Experience

- Illustrate and send dictated thank-you notes to the people who provided boxes.
- Ask the children to bring in smaller boxes to use in the various centers in your room.
- Decorate a large carton to make a present for another class.

# SHE'S COLORING ON MY PAPER!

*While several of your children are coloring at the art table, a girl reaches over and begins to color on a boy's paper. The second child howls, and the other children at the table are upset. What can the children do?*

| Children's Possible Solutions | Your Role |
|---|---|
| 1. The boy and some of the other children want the girl to be banished from the table. | Help the boy express his feelings and tell the girl why he is upset. Help all the children understand that in this safe classroom, we try not to do things that hurt others' feelings. |
| 2. Several other children suggest that the boy color on the girl's paper to show her how it feels. The girl replies, "I don't care." | Discuss whether hurting another's feelings will solve the problem. |
| 3. One child brings a fresh sheet of paper for the boy to use to begin a new picture. | Thank the child for this positive action and for being a good friend. |

## Extending the Experience

- Recognize that in mixed-age groupings, younger children often draw on older children's papers as a form of admiration. Help older children discuss this.
- Draw designs on paper. Then pass the papers around, letting everyone add details to each drawing. Display the collaborative creations for everyone to admire.
- Make a cooperative mural to go along with a theme the children are studying.

# THAT'S UGLY

*Your children are drawing and coloring at the art table. One child looks at a younger child's paper and says, "That's ugly." The younger child responds with tears. What can the other children do?*

| Children's Possible Solutions | Your Role |
| --- | --- |
| **1.** A girl says, "It is *not* ugly," and an older boy says gently, "He's just a little kid." The other children look on and say nothing. | Facilitate discussion. Ask, "How might this child feel? How did you color when you were younger?" Help your children understand that we are all beginners sometime. |
| **2.** A girl says, "I like your picture. Let's color together and make a book." | Compliment the girl on being a good friend and on recognizing how hard the younger child worked on his picture. |

## Extending the Experience

- View pictures of modern art with its drips and squiggles. Talk about how famous artists created the pictures.
- Visit an art museum to view modern art. Which paintings do the children like? Which do they not like? Why? Which paintings do they think that they could paint?
- Hang a gallery of artwork by children of all ages in the center's lobby or hallway where all works can be viewed and respected.

**Skills Used**
Social-Emotional Skills
Motor Skills

# HE'S NOT SHARING!

*Your children are at the art table working on self-directed activities. One child has most of the felt tip markers and refuses to share them. "I need these," he says.*

| Children's Possible Solutions | Your Role |
|---|---|
| **1.** One child begins to color with crayons. Another child uses art pastels. | Facilitate discussion with your children. Encourage them to look for a solution that is fair for everyone. |
| **2.** One child suggests a rule: "If you don't share, you have to leave the art table." Another child suggests a different rule: "You can only take one marker at a time. You put it back before you take another one." | Discuss the suggested rules, weighing all opinions equally. Help your children come to a fair decision. |

## Extending the Experience

- Plan an art activity in which the children sit together in pairs and each pair shares a small basket of supplies.
- Role-play scenarios in which one or more children won't share supplies or equipment. Discuss how each person in the scenario feels.

# STOPPED-UP GLUE BOTTLES

*The nozzles of several plastic glue bottles are stopped up. Your children squeeze and squeeze, but no glue comes out. How can the nozzles be opened?*

| Children's Possible Solutions | Your Role |
| --- | --- |
| **1.** One child pulls and twists and turns a nozzle cover until he pulls it off, exposing the insides. Now the glue comes out, but there is no way to control the amount or the direction. | Discuss what happened. Can anyone think of other ways to unstop the nozzles? |
| **2.** A girl suggests using scissors to pry off the dried glue around the tip of a nozzle. Another child suggests sticking something sharp down into the nozzle holes to keep them open. | Use the situation as an opportunity to discuss safety. "We use scissors just for cutting, and sharp, pointed things can hurt us." Suggest a safe alternative such as removing the glue from the nozzles with warm water. |

## Extending the Experience

- Together, plan an art activity that requires putting glue on paper in particular places, such as in the corners or around the edges.
- For some art projects, offer glue in small margarine tubs or film canisters and provide cotton swabs or coffee stirrers to use as applicators.

# DRIPPY PAINT

*A child is painting at the easel. The paint is a bit thin and is dripping down the paper. What can she do?*

| Children's Possible Solutions | Your Role |
|---|---|
| 1. The child gets a facial tissue and uses it to blot up the excess paint. | Acknowledge the child's resourcefulness. Make certain that lots of tissues are available. |
| 2. The child uses a facial tissue like a paintbrush to smear the drippy paint across the paper. | Compliment the child on learning a new painting technique. Ask if she would like to teach it to other children. |
| 3. Another child has been watching with interest. He thinks that the paint should be thicker. | Brainstorm with your children. What might they add to the paint to thicken it? Encourage all safe solutions and provide materials for the children to try. Offer small amounts of paint to experiment with. |

## Extending the Experience

- Drip small spoonfuls of paint on pieces of paper. Then tilt the papers to let the drips create colorful designs.
- Use paints of various thicknesses at the easel.
- View pictures of modern art with its drippy, splashed, and smeared techniques.

# NO PAINT-BRUSHES

*Your children want to paint at the easel, but somebody forgot to put out the paintbrushes. What can the children use to make their pictures?*

| Children's Possible Solutions | Your Role |
| --- | --- |
| 1. One child decides to use her fingers as a brush. | Support the child's idea. Provide soap and water for cleanup. |
| 2. Another child suggests using cotton swabs as paint-brushes. | Acknowledge the child's creative idea and provide cotton swabs for testing. |
| 3. A third child suggests bringing felt tip markers to the easel and using them instead of paint and brushes. | Compliment the child's problem-solving idea. Discuss how drawing with markers at the easel is different from drawing with them at a table. |

## Extending the Experience

- Use various items from around the room to apply paint to paper.
- Use items from nature, such as twigs, leaves, or dried grass, as paintbrushes. Discuss the results.
- Hang mural paper on an outside wall. Apply paint with various kinds of old, clean brushes such as toothbrushes, scrub brushes, bottle brushes, and hairbrushes. Talk about which brushes made which designs.

# WHY DOESN'T THIS STUFF STICK?

*Several of your children have used paste in making their collages. As the paste dries, many of the attached materials are falling off. What can the children do about this?*

| Children's Possible Solutions | Your Role |
|---|---|
| **1.** One child applies more and more paste to his creation. Some of the collage materials seem buried in paste. | Acknowledge the child's persistence. |
| **2.** Another child applies a thick layer of glue on top of the dried paste and reattaches her collage materials. | Provide materials and encouragement. |
| **3.** A third child removes all of the collage materials remaining on his paper and places his paper and the materials in a resealable plastic bag to take home. | Compliment the child's creative approach to taking his items home. |

## Extending the Experience

- In separate margarine tubs, mix paste by combining flour and water. Compare and discuss the various concoctions. Which are thicker? Thinner? Which hold better?
- Compare flour and water paste with commercial paste. How are they the same? How are they different?
- Experiment with other ways to stick things together. Explore adhesive bandages, strips of prepasted wallpaper dipped in water, various tapes, rubber cement, gummed labels, and glue.

# MIXING COLORS

*Two of your children are painting at the easel. They discover that where the red paint and the yellow paint run together, the color becomes orange. All the children are excited. How can they make other new colors?*

| Children's Possible Solutions | Your Role |
| --- | --- |
| **1.** Your children decide to mix other paint colors together, two at a time. | Compliment the children on their ingenuity and supply the needed materials. |
| **2.** One child suggests mixing all the paint colors together. | Encourage the children to name the new colors they create. |
| **3.** Another child suggests mixing colors by using one color of crayon over another. | Provide crayons for experimentation. |

## Extending the Experience

- Use plastic dropper bottles of food coloring to experiment with color mixing at the water table.
- Mix small amounts of different-colored tempera paint on squares of white construction paper and name the colors. Gather the papers together to make a color book.
- Look through pieces cut from colored, translucent file folders. What do the children see? What do they see when they look through two or more colors at once?

# THE MODELING DOUGH IS ICKY

*The modeling dough is drying out. Its texture is uneven, and when your children use it, the dough crumbles in their hands. What can the children do?*

| Children's Possible Solutions | Your Role |
|---|---|
| **1.** One child says that if you add water to a sponge, it gets soft. He suggests adding water to the modeling dough. | Support the activity and have your children discuss what happens. |
| **2.** Another child remembers that the dough needs to be kept in a covered container. She suggests putting a sign on the dough container: "Shut this tight." | Provide paper and felt tip markers for making the sign, and tape for attaching the sign to the container. |
| **3.** Several children remember making modeling dough in the past. They suggest making new dough. | Let each interested child make his or her own no-cook modeling dough. Then make a batch of fresh dough for the room. |

## Extending the Experience

- Discover and discuss different ways to work with modeling dough, such as pounding, patting, rolling, and ripping. Record the different actions on a chart.
- Use various implements, such as plastic knives, potato mashers, or garlic presses, with modeling dough.

# DRIED-OUT MARKERS

*Your children want to use felt tip markers for an art project, but they discover that several of the markers are dried out. What can the children do?*

| Children's Possible Solutions | Your Role |
|---|---|
| **1.** One child suggests brushing water over the paper and then drawing on the wet paper with the markers. | Encourage your children to try this solution. Provide materials and discuss the results. |
| **2.** Several other children suggest soaking the markers in a tub of water. | Encourage the children to try this solution and discuss the results. If the markers do work after soaking, are the colors as bright as before? How long do the markers continue to work? |
| **3.** A girl tries coloring with crayons and brushing water over her paper to make the designs look like they were made with markers. | Encourage the child's creative solution. Discuss the results. |
| **4.** Another child says, "Someone left the tops off the markers. We shouldn't do that!" | Discuss the consequences of not replacing the marker caps at the end of art projects. Why is it important to care for materials? |

## Extending the Experience

- Provide new felt tip markers and celebrate their bright, vibrant colors.
- Use watercolor pencils to make designs on paper. Then paint over the designs with water to create a watercolor look.
- Paint over construction paper with buttermilk and then draw on the paper with art pastels.

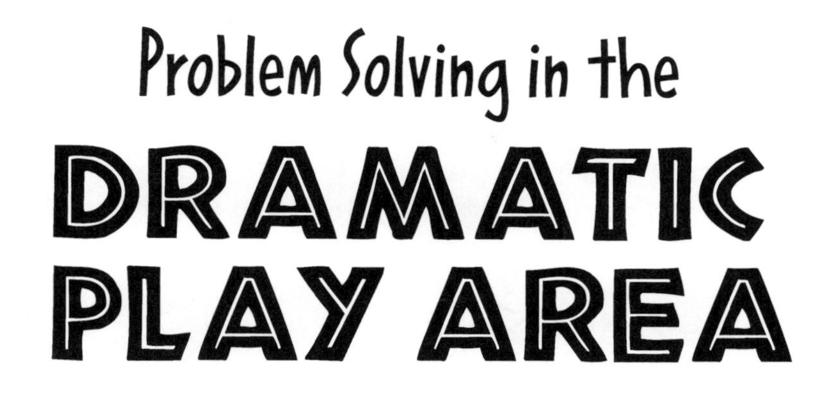
Problem Solving in the
DRAMATIC
PLAY AREA

# A SUPER SUPERMARKET

*Your children create a supermarket with empty containers of grocery items and toy plastic food. Because the children throw the objects in a heap on the shelves, the store is very messy and not much fun to play in. They want to play supermarket and attract new customers. How can they do this?*

| Children's Possible Solutions | Your Role |
|---|---|
| **1.** One child puts on an apron and announces she is going to put the food on the shelves. She wants to organize the items by categories (bread, fruit) to make them easier to find. | Compliment the child for relating her keen observations during visits to the supermarket. Encourage the group to discuss, then place items by classifications. |
| **2.** Someone suggests expanding the store and adding more shelves for aisles by using big blocks and boards. | Monitor the construction process to ensure the shelving is stable and won't topple. Talk about how more room will make the store easier to shop in. |
| **3.** Several children want to attract new customers by drawing colorful sale posters. | Provide markers, large sheets of paper, and tape to facilitate this idea. |

## Extending the Experience

- Send a letter to parents requesting exciting props for the newly expanded store—paper or canvas shopping bags, plastic flowers, used greeting cards, etc.
- Vote on a catchy name, such as "Zingo's Shop & Bag," for the supermarket's grand opening.
- Add accessories to create interesting roles (a cash register for a checker, a muffin tin and apron for a baker, and so on).

**Skills Used**
Motor Skills
Critical Thinking Skills

# THE BIRTHDAY CAKE

*The children are vigorously mixing together ingredients to make modeling dough. They plan on using the dough to create a cake for their doll's birthday party. However, in their exuberance, the water splashes on the floor and the flour flies all over everybody. They don't want the flour on their clothes. What can they do to prevent this?*

| Children's Possible Solutions | Your Role |
|---|---|
| 1. One baker suggests using bigger mixing bowls and then stirring the dough more gently. | Provide the necessary equipment. |
| 2. Another child recommends wearing aprons. | Compliment the child on remembering a way to keep her clothes clean. |
| 3. A girl thinks that everyone should remember their party manners and be more careful. | Discuss the girl's insightfulness about how we sometimes act differently in different situations. |

## Extending the Experience

- Provide accessories to prompt imaginative birthday play: candles for a cake, paper party plates, plastic flowers, wrapping paper, ribbons, and old greeting cards.
- Sing "Happy Birthday" to the doll.
- Create other pretend food items with modeling dough. Dry the dough in a warm oven. Paint the dough foods, then use them in the dramatic play area.

# ASK THE BOSS

*Drawn to the exciting truck play in the sandbox, a child asks to be a part of the action. The child who is directing the activity, the "boss," tells the boy he can't play. The boy really wants to join in. What can he do?*

| Children's Possible Solutions | Your Role |
|---|---|
| **1.** The boy tells the boss again, "I really want to drive a truck." | Give him moral support with your physical presence while the boss considers his request. |
| **2.** "Maybe I could dig out that dump truck that's stuck in the sand," the boy says. The boss considers his request and relents. | Provide some small digging tools (spoon, tongue depressor, etc.). |
| **3.** The boy offers to make his own truck to bring to the action. The others say he can join them when he's been properly trained in the use of the new vehicle. | Discuss an appropriate vehicle. Brainstorm how to create one with interlocking blocks. |

## Extending the Experience

- During sand and water play, encourage the children to build dams and bridges.
- Using the outdoor riding toys, involve the children in other leader-follower types of dramatic play, such as a bus driver and passengers, a traffic cop and drivers, and crossing guard and bicyclists.

# 1, 2, 3–HIKE!

*As your children become caught up in the escalating action of a football game, one of the players is hit by an airborne rubber ball. Although she is not hurt, she is now anxious about returning to the game. How can she feel at ease and still be a part of this sporting event?*

| Children's Possible Solutions | Your Role |
|---|---|
| **1.** A child suggests having boundaries and yard lines. She could be a spectator and safely cheer behind the lines. | Encourage the players to think about other safety rules that would benefit all participants. |
| **2.** Someone remembers seeing on TV how the coaches talk about important plays with walkie-talkies. | Discuss how exciting it is to use ideas seen elsewhere to think about new solutions. Help create a communication device with string stretched between two cans. |
| **3.** Another child says her brother played in the band during the half-time show. | Provide instruments for musicians and batons for drum majors. |

## Extending the Experience

- Explore these math concepts: call numbers for plays ("1, 2, 3—hike"), keep score on a scoreboard, make numbers for uniforms.
- Provide an assortment of safe, child-size balls to encourage playing other sports (plastic-foam baseball, basketball, soccer ball, etc.).
- Role-play what happens when a player is injured: coaches and trainers evaluate the injury; a stretcher might be used to carry the player off the field. Provide props such as ice packs, elastic bandages, and child-size crutches.

# THE BANK HEIST

*Several children have set up a bank. While they are banking, some robbers break in and steal the money. How can the tellers get the money back?*

| Children's Possible Solutions | Your Role |
|---|---|
| 1. One child tells the robbers to give up the money because they'll get caught if they try to spend it. | Compliment her on pointing out the consequences of their actions. |
| 2. Another child suggests calling the police. He wants the police to arrive with a loud siren and chase the robbers to get the money back. | Ascertain if there is sufficient, uncluttered space indoors for a chase to occur safely. Or let them continue their chase during outdoor play. |

## Extending the Experience

- Discuss ways to improve the bank's security system. Make a video camera with a round oatmeal box.
- Paint medium-sized boxes and hang them over children's shoulders with yarn to create a fleet of police cars.
- Provide opportunities for additional chasing activities outdoors—games of tag, nets for catching butterflies, etc.

# BEDTIME BATH FOR BABY

*Two children are caring for their baby doll. The doll is crying because he knows that after his bath, he has to take a nap. How can the parents get their baby doll to stop crying so he will go to sleep?*

| Children's Possible Solutions | Your Role |
| --- | --- |
| **1.** The mommy thinks it would be nice to sing the baby a lullaby and rock her baby to sleep. | Discuss how these actions seem to soothe and calm a fussy baby. |
| **2.** The daddy suggests giving the baby a bottle and then helping him burp up his air bubbles so his tummy won't hurt. | Compliment the child on his sensitivity to his baby's needs. Provide feeding-related props. |
| **3.** Both parents say they should read the baby a bedtime story—just like their own moms and dads do with them. | Help the children find a picture book to read. Comment on their awareness that certain rituals can be very comforting. |

## Extending the Experience

- Provide dramatic play props for other daily rituals in a baby's life: diapers, booties, hats, plastic feeding dish, spoons, bottles, a doll highchair, and so on.
- Enjoy other water play activities outdoors—wash doll clothes, float boats, wash tricycles, etc.
- Make up a song about giving a baby a bath.

# I HAD IT FIRST!

*Several children are playing shoe store. One clerk tells another, "I had the shoe sizer first!" The second clerk tries to grab it. The child with the measuring device screams, "It's mine!" What can the other children do?*

| Children's Possible Solutions | Your Role |
| --- | --- |
| **1.** Several customers want the clerks removed from the shoe store before they scare or hurt somebody. | Acknowledge the customers' feelings. Ask the bickering children to reflect on how others are feeling about their actions. |
| **2.** Others say the clerks should talk about sharing instead of screaming at each other. | Help the children develop verbal negotiating skills for a more peaceful solution. |
| **3.** One child thinks that they should have more measurers. Maybe they could make some. | Listen to ideas for creating this prop, then supply the necessary materials. |

## Extending the Experience

- Play a classification game. Sort the shoes and stock the store shelves by color, size, fasteners, etc.
- Sing an action song about shoes, such as "One, Two, Buckle My Shoe."
- During transition times, creatively move between activities wearing different types of imaginary shoes, such as tiptoeing with ballet shoes or galloping in cowboy boots.

# BUNNY'S BROKEN LEG

*Several children built a rabbit hole out of boxes. When they dropped their stuffed bunny into the hole, he came out the other end with his leg all twisted. The children decide that the bunny's leg is broken. They want to help their bunny feel better. How can they do this?*

| Children's Possible Solutions | Your Role |
| --- | --- |
| 1. A few propose bringing him to the veterinarian to get patched up. | Applaud the children's decision to get quick medical attention for his injury. They have made a sound decision. |
| 2. Some think they should bandage his leg to keep it straight on the trip to the vet. | Help provide a first-aid kit so the children can provide temporary assistance for the ailing rabbit. |
| 3. Another says the vet should give him a shot in case of infection. However, they can pat the bunny so he won't be afraid. | Discuss how important nonverbal support can be during an emergency. |

## Extending the Experience

- Collect props (tongue depressors, gauze, a white shirt for the vet's lab coat, a stethoscope, etc.) for a veterinarian's animal clinic.
- Have your children dictate stories about their own injuries. Let them decorate their stories with crayon designs and adhesive bandages.

# I WANT TO BE THE BRIDE

*The children decorate for a mock wedding. Two girls begin to argue. There is only one white veil, and they both want to be the bride. What are they going to do?*

| Children's Possible Solutions | Your Role |
| --- | --- |
| **1.** Holding up pieces of colorful, lacy material, a child suggests that one girl could design a beautiful bridesmaid's dress. | Listen attentively while the girls explain why this idea will or won't work. |
| **2.** A child relates that at her aunt's wedding, the lady who caught the bouquet was the next bride. Maybe they could take turns. | Explain that this is a wedding tradition. See if the two girls are flexible enough to try out this custom. |
| **3.** The "florist" tells the girls to forget about the veil and wear flowers in their hair. | Offer the wedding party colorful tissue paper and pipe cleaners to create lovely headpieces and bouquets. |
| **4.** One girl suggests that they both be brides at a "double wedding." | Discuss who might be brides in a double wedding: Two sisters? Two friends? Who else? |

## Extending the Experience

- Create elegant wedding invitations with paper, markers, glue, and glitter.
- Make a cake from a mix. Frost the cake and add sprinkles to make a beautiful wedding cake to serve at the "reception."
- Take photos for a wedding album. Place the album in the library to share long after the wedding is over.

# BOYS CAN'T BE SECRETARIES

*A typewriter, a desk, and a file cabinet are set up for an office center, and several girls are playing there. A boy joins the play. One of the girls teases him and says, "You're stupid. Boys can't be secretaries!" How can the boy and other children respond to her?*

| Children's Possible Solutions | Your Role |
|---|---|
| 1. The boy says, "It makes me mad when you tell me I can't be a secretary! Why not?" | Compliment the boy on expressing his feelings. Ask the girl to clarify the problem by answering the boy's question: Why can't he be a secretary? Help them reach an agreement. |
| 2. Another child says the boy can too play. Her dad uses a typewriter and a computer at work. He's an executive secretary. | Talk about how office equipment can be used by both men and women and how offices have both men and women working together in them. |
| 3. Several others mention various roles (office manager, clerk, bookkeeper, etc.) to add to their new office center so everyone can play. | Comment on the way cooperation makes an office run smoothly. |

## Extending the Experience

- Discuss being respectful of others. Practice expressing feelings without calling names.
- Set up a typewriter or a computer to write a class newsletter.
- Place a phone in the office center for practicing telephone etiquette. Add a pad and paper for taking phone messages.
- Visit a real office with both men and women in it, and talk to the various workers about their jobs.

# THE HAIR SALON

*Several children have set up a hair salon. One of the stylists is waiting for another stylist to finish using the toy hair dryer on a customer. She doesn't like waiting for the equipment. What can she do?*

| Children's Possible Solutions | Your Role |
|---|---|
| 1. The girl goes to the art center and makes a new "hair dryer" so she can style hair right away. | Compliment the child on thinking of a way to use art center materials in the dramatic play area. |
| 2. Several customers say the stylists should have appointment times so they won't have to wait. | Comment on the children's efficient idea. Supply an appointment book, a pencil, and a telephone. |
| 3. The girl asks the other child if she will trade the blow dryer for the shampoo cape. | Support this negotiation strategy. Make sure a variety of props are accessible. |

## Extending the Experience

- Add a manicure table to the hair salon. Be sure to include cotton swabs for brushes, craft sticks for nail files, and a dish for soaking fingers.
- Display beauty magazines or hairstyle magazines. Children can match styles (bangs, long, curly) to their own. Create a class graph of hair colors or styles.
- At the art center, have children curl paper strips with their fingers to design individual wigs.

**Skills Used**
Critical Thinking Skills
Social-Emotional Skills

# A YARD SALE

*Looking through several bags of books, games, toys, and dress-up clothes donated by their parents, the excited children decide to set up a pretend yard sale. How can they decide what to sell?*

| Children's Possible Solutions | Your Role |
|---|---|
| **1.** One child suggests selling things if you don't use them anymore or are tired of them. | Discuss how interests change as children grow. |
| **2.** Other children say they should sell the items that are too small for them now. | Talk about how children's bodies change, too, as they grow. |
| **3.** A boy thinks everybody should pick one item from each category to put in the sale. | Help the children organize a manageable system for their selections. |

## Extending the Experience

- Discuss prices. Write numbers on tags and put them on the sale items.
- Make flyers advertising the yard sale.
- Provide an assortment of newspapers, bags, and boxes for your children to use to wrap up and carry newfound bargains.
- Let the children take turns buying and selling items at their "garage sale."

# FIRE! FIRE!

*A boy is baby-sitting the baby dolls in the housekeeping center. While he is making lunch, he pretends that the food catches on fire. He yells for help. What can the other children do?*

| Children's Possible Solutions | Your Role |
|---|---|
| 1. One child who is passing by grabs a block and uses it as a fire extinguisher. | Acknowledge the child's connection between "seeing" the fire and putting it out. Have your children consider making a fire extinguisher to add to the props in the housekeeping center. |
| 2. Another child calls the emergency number—911—on the phone in the housekeeping center. She tells the operator where the house is located. | Prompt the child to clarify any details for the operator. Compliment her on her 911 skills. |
| 3. The babysitter gets the baby dolls out of the house quickly. | Discuss why it is important to escape from the fire and not try to hide from it. |

## Extending the Experience

- Supply fire helmets, hoses, a steering wheel, and large blocks. Let your children make a fire truck.
- Practice fire safety by reacting to the command "Stop, drop, and roll" during transition times.
- Visit your neighborhood fire station to see the trucks and equipment. Talk to the firefighters about fire safety.
- Review fire safety information. Have a fire drill.

# FINDING CINDERELLA

*After hearing their favorite royal fairy tale, the children get out their puppets and create a stage using a cardboard box. The prince puppet then asks, "What if Cinderella doesn't lose her shoe? How will I find her?"*

| Children's Possible Solutions | Your Role |
|---|---|
| 1. Several children suggest a password that Cinderella could whisper in the prince's ear as they dance. | Encourage the children to think of an appropriate royal password. |
| 2. One girl says that Cinderella should give the prince a secret ring with her picture inside. | Provide exciting collage materials for the children to use to create a miniature ring. |
| 3. A computer-savvy child recommends that Cinderella give the prince her e-mail address so they can get in touch by computer. | Provide materials to "write" the e-mail address. Give the children small boxes to use as computers. |

## Extending the Experience

- Play dreamy waltz music so that Cinderella and the prince can dance together all night.
- Paint boxes gold and sprinkle them with glitter. Arrange the boxes to make a palace dollhouse where the prince and the new princess will live happily ever after.

**Skills Used**
Creative Thinking Skills
Social-Emotional Skills

# SUNKEN TREASURE

*Your children are on an underwater adventure. They discover sunken treasure in a shipwreck. Suddenly, they are surrounded by howling sea monsters. How can they rescue the treasure chest and escape the monsters?*

| Children's Possible Solutions | Your Role |
| --- | --- |
| **1.** One child suggests dropping a strong net between the divers with the treasure chest and the monsters to separate them. | Clear an area on the ship's deck for energetic sailors to work fast. Provide a prop such as a blanket for a net. |
| **2.** Some of the sailors on the ship suggest pulling up the treasure chest with a giant magnet, leaving the monsters behind. | Challenge the children's thinking. Ask them: Will a magnet pick up gold, silver, or jewels? |
| **3.** Another child says they should use loud sounds to chase the monsters away. When the monsters leave, they can go rescue the treasure chest. | Encourage trying this series of options. Acknowledge the child's patience. |

## Extending the Experience

- Design exquisite bracelets and crowns from strips of foil wallpaper. Glue on glitter, sequins, and beads.
- Create an underwater viewer by cutting the top and bottom off a cardboard milk carton. Cover the bottom of the carton with plastic wrap and secure it with a rubber band. Use it to look for jewels in the water table.
- For more high seas adventures provide sailor hats, life vests, and plastic paddles. Sing sea chants while sailing.

# TRIP TO DINOSAUR WORLD

*The children pretend they have just won a free trip to Dinosaur World; however, they must arrive there by midnight to collect their prize-winning tickets. How can they get there on time?*

| Children's Possible Solutions | Your Role |
| --- | --- |
| **1.** One child suggests all the children pretend to be robots that know how to pack suitcases extra fast. | Provide an assortment of luggage to choose from. Discuss how robots move. |
| **2.** Other children suggest that they all hop on a super-speedy, magic airplane that will get them to Dinosaur World super fast! | Focus their attention on specifics: What does the airplane look like? What are the seating arrangements? Is there enough room for everyone and their suitcases? How long will the trip take? |

## Extending the Experience

- Construct a new ride for Dinosaur World using the climbing apparatus. Decorate it with colorful crepe-paper streamers.
- Have refrigerator cartons and hollow blocks available so your children can build a hotel to stay in while they visit Dinosaur World.

# DRIVE-UP PICNIC

*The children are riding their trikes on the playground. They decide it is time for dinner and they want something to "eat." What can they do?*

| Children's Possible Solutions | Your Role |
|---|---|
| **1.** Some children make a menu board for a fast-food drive-through restaurant. The drivers take turns reading the menu and ordering some play food. Another child pretends to put on a headset, take orders, ring up money, and bag food. | Provide art materials. (Store some basic art supplies outdoors in a dry but accessible spot for future creating.) Help the children brainstorm other interactions between customers and drive-through workers. |
| **2.** One trike rider says, "I have a picnic in my trunk. Let's go eat over by the tree." | Provide a blanket for the children to sit on. Join in the fun! |

## Extending the Experience

- Extend the restaurant theme indoors with additional roles and props: chefs (tall hats and spatulas), wait staff (adding machine, tape, plastic food), etc.
- Discuss favorite foods. Make human graphs—line up after naming favorite sandwiches (peanut butter, cheese, turkey) or ice cream flavors (vanilla, strawberry, chocolate). Which ones are the favorites?

Skills Used
Creative Thinking Skills
Critical Thinking Skills

# THE HEALTH CLUB

*The children find some long boards near the climbing bars. They want to use them to make a health club. What can they do?*

| Children's Possible Solutions | Your Role |
|---|---|
| **1.** One child says, "They look like the bench you lie down on to pump iron." | Reinforce the children's interest in fitness. Encourage them to think of safe objects to use as weights. |
| **2.** Two children suggest sitting on them like a rowboat and making a pretend rowing machine. | Comment on the children's collaborative effort as they row together. |
| **3.** Others place the boards on the ground and walk on them. "Look! We've made a walking track!" | Support the children's enthusiasm in this positive project. |

## Extending the Experience

- Create additional gym equipment out of dowels, boxes, aluminum foil, etc.
- Design membership cards for the new health club.
- Exercise to peppy music. Afterwards, listen to one another's heartbeats with a stethoscope. Discuss the connection between exercise and a healthy heart.

# Problem Solving with

# MANIPULATIVES

# STRINGING BEADS

*Your children are stringing beads onto long laces. They decide they want to make necklaces. How can they do this?*

| Children's Possible Solutions | Your Role |
| --- | --- |
| **1.** A child picks up one end of his string of beads. As he lifts the string, the beads slide off the other end. He wraps many layers of masking tape around the far end of the string, making it so thick that the beads cannot slip off. Now he strings beads again, then brings the string to you to tie into a necklace. | Compliment the child on his problem-solving technique. |
| **2.** Another child has two strings knotted together, and the knot holds her beads on the first string. When she holds the ends together around her neck, the necklace is much too long and one string has no beads on it. | Watch and wait to see if the child wants help. Encourage her attempts to discover a solution. |

## Extending the Experience

- Ask parents to donate used shoelaces to use in stringing activities. Challenge your children to find items to string. Old keys work well.
- Dye stringable pasta by placing 1 cup of pasta in a plastic jar with 1 tablespoon of rubbing alcohol and 1 teaspoon of food coloring. Screw on the lid and shake well. Spread the pasta pieces on waxed paper to dry. Let your children string them on shoelaces or yarn.

**Skills Used**
Motor Skills
Critical Thinking Skills
Creative Thinking Skills

# STIFF PAPER PUNCHES

*Your children are using the paper punches to make sewing cards. Two of the punches are stiff and difficult to work. What can the children do?*

| Children's Possible Solutions | Your Role |
| --- | --- |
| **1.** Several children take their punches to the sink and run water on them. "The punches feel too dry," they explain. | Encourage the children's efforts and compliment them on their idea. Discuss the results. |
| **2.** One child has seen the teacher rub petroleum jelly on the insides of paint container lids to keep them from sticking to the containers. She wants to rub petroleum jelly on the hole punch. | Provide materials and support the child's efforts. Discuss the results. What are the advantages and disadvantages of this method? |
| **3.** Another child has seen his mom squirt oil on things (sewing machine parts, a stiff door hinge) to make them move smoothly. He wants an oil can so he can put a drop of oil in the punches' joints. | Compliment the child for applying a workable idea that he observed elsewhere. Supply the oil if you have it. If not, brainstorm to see if you have a possible substitute. |

## Extending the Experience

- Show your children how oil acts as a lubricant to help metal parts move smoothly over one another. Together, find a squeaky hinge to oil.
- Make collages with the paper punchouts.

# JAR LIDS AND BOTTLE TOPS

*Your children have collected jar lids and bottle tops to see how many it would take to fill a special box. The whole class has counted 246 lids and tops. Now, what can they do with these?*

| Children's Possible Solutions | Your Role |
| --- | --- |
| **1.** Some children take a pile of lids and separate them into two stacks, one for rough lids and one for smooth ones. They run their fingers over the raised lettering on some of the lids to show you where they are "rough." | Compliment the children on their thinking. Ask if there are other ways they might group the lids. |
| **2.** One girl sorts some of the lids by color. She gets a container of teddy bear counters and puts a matching-colored teddy bear on each colored lid. She looks for a long time at the metallic lids, then she smiles and jumps up to get a container of keys. She puts a matching-colored metal key onto each lid. | Congratulate the child on her colorful solution. |
| **3.** Two children lay tops and lids in two parallel lines that curve about, leaving enough room between the lines for a small car to roll through. "We're making a road," one of them says. | Admire the children's work, and if they want to add intersections and side roads, discuss ways they might do so. |

## Extending the Experience

- Find rough and smooth items and place them in separate piles.
- Explore with several colors of teddy bear counters and matching-colored strips of paper.

# LOTS OF LETTERS

*Several children have gathered wooden, plastic, and magnetic letters, as well as the letter cards from an alphabet game. They would like to play a game with all these letters. What game can they make up?*

| Children's Possible Solutions | Your Role |
| --- | --- |
| **1.** Two children lay out the letter cards and bring objects from the room to put on each letter. "Bring a truck for *T*," one child says. | Admire the children's work. If they cannot think of items for some letters, suggest they look in ABC books for ideas. |
| **2.** One child takes wooden letters to the art center. He traces the letters of his name, cuts them out, and pastes them to a sheet of construction paper. | Provide materials for children who want to copy the child's idea. |
| **3.** Some children put the magnetic letters on the side of a metal file cabinet. They sing the alphabet song, finding the letters as they go, putting them in order on the cabinet. | Watch quietly. |

## Extending the Experience

- Dip sponge letters onto paint pads and then press them onto paper to make letter prints.
- Walk, hop, or skip on large letters outlined on the floor with masking tape or drawn in the dirt outdoors.
- Fill the library area with a variety of ABC books.

# TOO MANY CARDS

*Several children are playing with a deck of cards, drawing from each other's hands and matching pairs. They are having trouble holding their cards. What can they do?*

| Children's Possible Solutions | Your Role |
| --- | --- |
| **1.** One child lays her cards faceup on the floor. Now she can see all of them, but so can the other children. Suddenly, she is losing all of her cards as the others choose exactly what they want. | Compliment the child on her solution. Brainstorm ways she could hide her cards from others. |
| **2.** Another child also lays his cards faceup on the floor, but first, he stands two tall books on end so they hide his cards. When someone wants to select a card, he picks up his cards, holds them out, and then, after the selection has been made, spreads them out behind the books again. | Comment on the child's creative thinking. Ask what the others think of this idea. |
| **3.** A girl takes two clothespins from the art center. She divides her cards into two groups and fans out each group. She clips a clothespin onto each group. When other players point to the cards they want, she removes them from the clothespins. | Praise the child's problem-solving skills. |

## Extending the Experience

- Arrange numbered cards in order from 1 to 10.
- When two or more decks of playing cards get mixed together, have your children sort them.
- Make up new games using old playing cards.

# RAINBOW PEGS

*Several children are playing with pegboards and pegs. A girl is putting "candles" on a cake. One of the boys is making a rainbow. They both want more red pegs. What happens now?*

| Children's Possible Solutions | Your Role |
|---|---|
| **1.** The boy cries, "Give me the red pegs. The girl refuses to relinquish any red pegs. The boy whines, "Teacher, make her give me her red pegs." | Watch carefully to see what occurs. Be ready to step in if needed. |
| **2.** The girl counts the empty holes in the boy's row of red pegs. There are four. She looks at his other pegs and announces, "I'll give you four red pegs if you give me six purple pegs and three green pegs. I need them for my birthday cake." The boy checks his rainbow to see if he needs those other pegs, then agrees to the trade. | Praise the children for their peaceful solution. |

## Extending the Experience

- Explore with modeling dough and pegs.
- Divide your children into pairs. Give each pair one pegboard and some pegs. Let them work together to create a design.

# NO THROWING ALLOWED

*While your children are playing with blocks, one child throws a block at another child. The other children don't like this. What can they do?*

| Children's Possible Solutions | Your Role |
|---|---|
| 1. One girl thinks that any toys that are thrown should be put in the closet so no one can play with them. | Ask what the other children think of this idea. Brainstorm other options. |
| 2. A boy suggests a rule: No throwing blocks at school. | Encourage discussion. Offer materials if the children want to make "Do not throw" signs to put in the block area. |
| 3. Other children say that if they want to throw things, they should go outside and throw balls. | Let your children go outside and throw balls if they want. |

## Extending the Experience

- Have a group meeting to go over classroom rules.
- Talk about where and what your children can throw. Let them demonstrate and practice this.
- Outside, toss balls through a Hula-Hoop hanging from a tree branch or a clothesline.

# THESE PEOPLE DON'T LOOK LIKE ME!

*Several children are playing with the dollhouse. One boy who uses a walker says, "Hey, there's no walker in this house." A girl who wears glasses notices that none of the toy people wear glasses. They want to change this. How?*

| Children's Possible Solutions | Your Role |
|---|---|
| **1.** The boy gets pipe cleaners from the art center and begins fashioning a walker. | Encourage the child's efforts. |
| **2.** The girl, who is blonde, takes a pen and begins drawing glasses on a dark-haired figure. "Hey," shouts a brunette girl, "That one is me. Put glasses on this one." She hands her a blonde figure. | Be ready to mediate disputes. Support both girls' efforts. Discuss other ways to add glasses besides writing on toys. |

## Extending the Experience

- Find a magazine picture to represent each child. Cut out the pictures, mount them on heavy paper, cover them with clear self-stick paper, and put them in the manipulatives toy area.
- Add a variety of multiethnic, differently-abled figures to your dollhouse. Have your children help you select these from school supply catalogs.

# THIS PUZZLE IS TOO HARD

*A boy has been sitting on the floor working with a tray puzzle that has many pieces. He is unable to complete it and is feeling frustrated. He doesn't want to work on the puzzle anymore. What can he do?*

| Children's Possible Solutions | Your Role |
|---|---|
| **1.** The boy angrily stacks the puzzle pieces on the tray and carries them to the shelf. A child sees him and tells him he can't put the puzzle away like that. | Do not intervene, but watch carefully to see that tempers do not get out of hand. |
| **2.** One girl notices his frustration and says, "I'll put it together for you." | Suggest that the girl show him *how* to put the puzzle together. As she works, describe her actions: "She puts the biggest pieces in first. She looks for a piece that is the same color as this." |

## Extending the Experience

- Glue large color copies of your children's photos to heavy paper. Cover each one with clear self-stick paper, cut it into puzzle pieces, and place it in a separate resealable bag. Children love putting together these "friendly" puzzles.
- Offer a variety of puzzles for varying skill levels to encourage successful puzzle experiences for everyone.

# NOT ENOUGH BUILDING LOGS

*Three children have been creating with the building logs. They decide to build a city. "We need more logs," they say. They look around the classroom. There are no more logs. What can they do?*

| Children's Possible Solutions | Your Role |
|---|---|
| **1.** One child suggests bringing some from home. | Brainstorm ways of marking and counting the logs the children bring from home so they will take home the same logs that they bring. |
| **2.** Some children think they should use cardboard tubes. They can tape them together. | Make these materials available. |
| **3.** Another child suggests asking other classrooms if they can borrow their building logs. | Allow the children to go to other classrooms and to ask the teachers if they can borrow their logs. Again, decide how to count and mark the logs so they can all be returned to the correct classrooms. |

## Extending the Experience

- Roll grocery bags lengthwise to make "logs" and secure the ends with masking tape. Fasten these together with glue, rubber bands, or masking tape to build structures.
- Team up with other teachers and share materials from room to room. One week you have all of the building logs and the other class has all of the interlocking blocks.

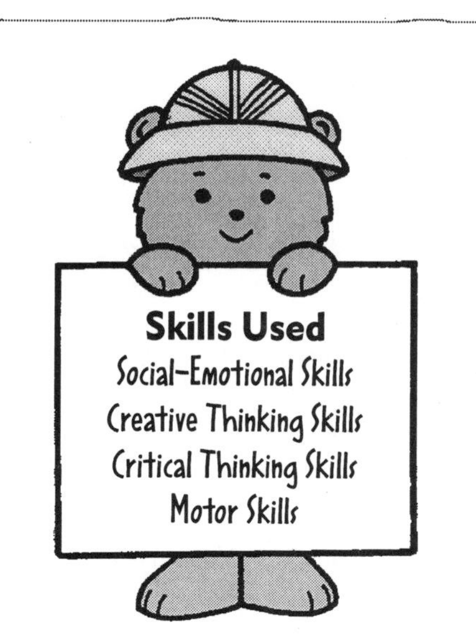

# JUMPING PAPER CLIPS

*Two children are playing with magnets and an assortment of paper clips. One boy is holding six disc magnets stacked together in one hand. As he passes his hand a few inches above the clips, a few of them seem to "jump" up to his hand. The children want to know why this is happening. What can they do?*

| Children's Possible Solutions | Your Role |
|---|---|
| **1.** Some children try passing their empty hands over the paper clips. Nothing happens. | Encourage discussion of the children's observations. What is different about the two situations? |
| **2.** The boy passes his "magnet" hand over the paper clips a second time, only this time his hand is farther away. Nothing happens. When he lowers his hand, the paper clips jump up again. | Support the child's actions. What does he think is happening? |
| **3.** Another child has been watching carefully. She passes a stack of ten disc magnets over the clips. Many clips jump high toward the magnets and several stick to them. She stacks all the remaining magnets together, passes them over the clips, and watches the clips jump even higher. | Praise the child's observations and logical thinking. |

## Extending the Experience

- Explore a variety of magnets including refrigerator magnets; small, strong magnets from a hardware store; horseshoe magnets; and so on.
- Place paper clips in a jar of water and screw on the lid. Provide magnets and encourage exploration.
- Place several small magnetic and nonmagnetic items in a shallow, plastic box with a clear lid. Move a strong magnet across the lid and observe the materials inside.

# OPPOSITES

*During circle time your children play a movement game about opposites, making themselves wide, then narrow; short, then tall; fast, then slow. They are excited about their discovery of opposites and want to explore this new concept during their free play time. How can they do this?*

| Children's Possible Solutions | Your Role |
| --- | --- |
| 1. One child paints at the easel, making a pattern of wide lines and narrow lines. | Provide materials. Offer to label the child's finished work. Encourage him to hang the creation where others can see it. |
| 2. Other children decide to start a collection of large and small items. They find a large paper clip and a small one, a child's shoe and a doll's shoe, a large wooden letter and a small magnetic letter, a new pencil and a pencil stub, and so on. | Compliment the children on their idea and their resourcefulness. Provide a place to display the collection and offer to help them label it. |

## Extending the Experience

- Show opposites by moving fast and slow, climbing up and down, and walking backward and forward.
- Take photos of your children walking up and down steps, standing right-side-up and hanging upside-down on the climbing bars, and standing near and far. Use the photos to make a book of opposites.

Skills Used
Critical Thinking Skills
Motor Skills
Creative Thinking Skills
Social-Emotional Skills

# WHO HAS MORE?

*Two children are playing with wooden unit cubes. Both children stack their cubes, and the stacks are the same height—nine cubes tall. They knock down their towers and one child's cubes scatter over a wider area than the other child's. The children want to know if they still have the same number of cubes. What can they do?*

| Children's Possible Solutions | Your Role |
|---|---|
| 1. Both children count their cubes. They each have nine. | Help the children with the counting, if necessary. |
| 2. One child says, "My blocks take up more room on the floor, so I must have more." | Encourage the children to lay the cubes on the floor side by side or ask them if they would like to compare their cubes by weighing them on the balance scale. |
| 3. Both children stack up their blocks again and compare the height of their towers. | Ask the children what they observe. Encourage them to count the cubes in each stack. Evaluate any chosen solutions with your children. |

## Extending the Experience

- Count and compare numbers of things.
- Divide your children into two equal groups. Count each group. Have them line up facing each other. Count both groups. Have one group spread out. Count both groups again. Discuss the results.

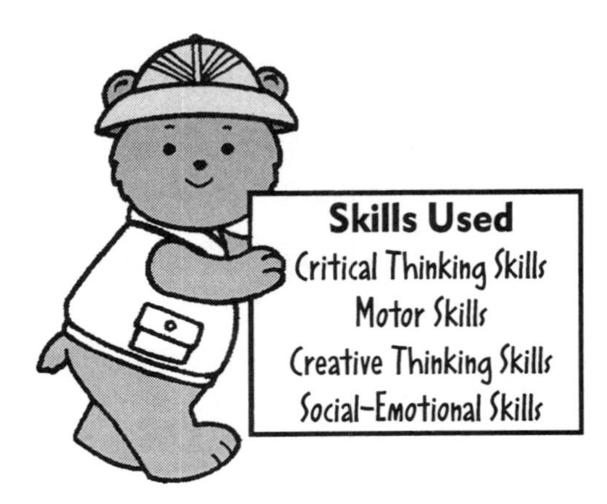

# THE SLANTED FLOOR

*When your children play with small balls or marbles indoors, the balls always roll in one direction on the classroom floor, ending up by one wall. How can they incorporate this into their play?*

| Children's Possible Solutions | Your Role |
| --- | --- |
| **1.** One child blocks the rolling marble with her hand and pushes it back to the starting point. When she lets go, the marble again rolls toward the wall. | Support the child's action. Discuss her results. |
| **2.** Some children get a shoebox and set it against the wall. They take turns trying to roll marbles into the box "target." | Provide materials. Praise the children for their creative idea. Encourage others who want to make targets from different materials. |

## Extending the Experience

- Brainstorm ways to make targets outside to use with balls.
- Use a rubber ball to test different areas of the play yard to see which ones are level.

# WHO LIVES ON THIS FARM?

*Several children are playing with a small farm set—a barn, various toy animals, and a toy "farmer" figure. The children want to have the farmer's children feed the animals, but there are no child figures. What can they do?*

| Children's Possible Solutions | Your Role |
|---|---|
| **1.** A girl takes a slot-type clothespin and felt tip markers from the art center. She draws a face and clothes on the clothespin. The clothespin child "feeds" the animals. | Compliment the girl on her creative solution. Support her decision to use art materials in the manipulatives toy area. |
| **2.** A boy makes "people" from spools. Another cuts pictures of children from a magazine and tapes them to blocks. | Provide needed materials and compliment the children on their creations. |

## Extending the Experience

- Sing "Old MacDonald Had a Farm," but instead of singing about the sounds the animals make, sing about the foods they eat.
- Read books about farms. Talk about what the various animals eat.
- Visit a farm, if possible. Ask if the children can help feed the animals.

**Skills Used**
Creative Thinking Skills
Motor Skills
Social-Emotional Skills

# PIRATE LIGHTS

*Some children are under a table, playing with flashlights. One child suggests they make a pirate ship and use the flashlights to send signals to other pirate ships. How can they do this?*

| Children's Possible Solutions | Your Role |
|---|---|
| 1. The children cover the table with a plastic tablecloth from the art area and a blanket from the dramatic play area. They hide in their new "ship." | Encourage the children's creative play by providing materials and by supporting their decision to use materials from more than one center. |
| 2. Two children go to the block area and build a block ship. | Join in the excitement of having pirate ships in the room. |
| 3. One child arranges two chairs for his ship and covers them with a transparent red scarf from the dramatic play area. When he flashes his signal, it looks red. | Point out the color of the child's signal light. Compliment him on his colorful idea. |

## Extending the Experience

- Turn out the classroom lights and use flashlights for reading, art, dramatic play, and other activities.
- Hold flashlights while doing movement activities. Enjoy the pattern of dancing lights.

# THE CIRCUS

*The circus is in town and many of your children have attended the performances. They want to create a circus in the manipulatives toy area. How can they do this?*

| Children's Possible Solutions | Your Role |
| --- | --- |
| **1.** They use blocks and other small manipulatives to build three rings for the circus. They put teddy bear counters and small plastic animals in the rings and use toy people as animal trainers. | Encourage the children's creative manipulations. Discuss the work of people and animals. |
| **2.** Two children want to make a trapeze. They build a standing crossbar from some building toys and hang a pipe cleaner trapeze from it. | Provide circus music to add an authentic atmosphere. |
| **3.** Some children want to make a circus tent. They place classroom chairs in a circle and drape a sheet over the chairs. They place zoo animals from the block area into shoebox cages around the outside of the tent. | Take photos of the finished creations inside and outside the tent. Include the creators in the photos. Encourage children who want to put these into a photo album along with dictated captions. |

## Extending the Experience

- Prepare popcorn and lemonade to "sell" at a "concession stand." Use art center supplies to make money for the concession stand.
- Let your children make up and star in their own circus. Provide materials for simple costumes or have them use props from the dramatic play area.

# BROKEN PUZZLES

*Some of the tray puzzles have pieces missing. The children think they aren't good for puzzles anymore, but they don't want to throw them away. What can the children do with them?*

| Children's Possible Solutions | Your Role |
|---|---|
| **1.** One child wants to fix one of the puzzles by drawing in the missing pieces. She shows how by putting in all the remaining pieces, and then coloring the empty spaces on the puzzle tray with a felt tip marker. | Support the child's effort. Supervise carefully while you allow her to color in missing spaces. |
| **2.** Another child suggests getting rid of all the puzzle pieces and using the empty trays as building platforms in the block area. | Lead the children in a discussion of this solution. |
| **3.** Other children collect all the full-figure pieces (a Papa Bear, three bowls, and a chair) and take them to the dollhouse to play with them there. | Compliment the children on their original thinking. |

## Extending the Experience

- Add new puzzles to the manipulatives area. Encourage discussion about caring for the puzzles.
- Make a puzzle by folding a sheet of paper in half and cutting out a shape through the fold. Open the paper to discover a "puzzle tray" and a cut-out puzzle piece.
- Find other toys with missing pieces. What can the children do with these?

Problem Solving in the

# BLOCK AREA

# GOLD SHIPMENT

*The children use a toy bulldozer and front-end loader to lift a shipment of "gold bars" (unit and half-unit blocks) into two "armored trucks" (toy dump trucks). A concerned "guard" wants to know if each truck is getting the same amount of gold. How can the children check this?*

| Children's Possible Solutions | Your Role |
|---|---|
| **1.** A child relates that the teller at the bank counts her mommy's money into piles. She suggests having tellers. | Discuss how exciting it is to use ideas seen elsewhere to think about new solutions. Help designate areas in which to sort the different sizes of bars. |
| **2.** A boy suggests laying down the gold bars end to end in two long lines on the floor so everyone can see that each line has the same amount. | Compliment the boy on using his keen observation powers. |
| **3.** Someone remembers seeing gold bars weighed on scales during a TV program. | Supply a pan balance scale. Discuss how many small blocks it takes to equal the weight of a large block. |

## Extending the Experience

- Set up a bank in the dramatic play center.
- Create dazzling gold bars by painting sanded wood scraps with gold paint or by painting them with glue and sprinkling them with gold glitter.
- Compare other items on the balance scale that weigh more than, less than, or the same as the gold bars.

# THE TALL TOPPLING TOWER

*The children try to build a skyscraper using various sizes of wooden blocks. Each time the tall tower almost reaches shoulder height, it falls down. The children are becoming frustrated with the unbalanced blocks. What can they do?*

| Children's Possible Solutions | Your Role |
| --- | --- |
| **1.** One child suggests watching to see if the tower falls down when a big block is placed on top or when a little block is placed on the top. | Compliment the child on the connection he has made between the size of the blocks and the way they balance. |
| **2.** Some children say that it is impossible to build a tower with more than a certain number of blocks. If they do, the tower will crash down. | Ask the children to predict how many blocks high a tower can be before it topples over. Have them test their prediction. |
| **3.** Another builder thinks they should build the tower against the wall because the blocks don't fall over when they are stacked there at cleanup time. | Expand on this idea. Show pictures of city buildings where outside walls are connected. |

## Extending the Experience

- Hang up posters of famous architectural structures to look at and discuss.
- Experiment with a variety of stacking materials, such as nesting boxes and plastic mixing bowls.
- Take a trip to a construction site to see how cinderblock foundations and brick walls are laid.

# LOOK-ALIKES

*Several children build a giant castle using large hollow blocks and decorate it with unit blocks. They decide they want each side of the castle to look identical. How can they do this?*

| Children's Possible Solutions | Your Role |
|---|---|
| **1.** One child tells the others, "How about playing Follow the Leader? I'll put up a block on one side, then you put up the same kind of block on the other side." | Make sure the children have lots of space to carry out this project. |
| **2.** Some children suggest there be one person who tells the others where to put which blocks. | Help the children identify key words (*ramp, cylinder*) to describe the blocks. You might want to introduce the word *symmetrical.* |

## Extending the Experience

- Create small buildings with table blocks. Hold unbreakable rectangular mirrors next to the structures to see symmetrical images reflected.
- Have one child close her eyes. Let a friend place a table block in each of her hands. Ask her to tell if the blocks are the same or different.

**Skills Used**
Motor Skills
Critical Thinking Skills

# A BROKEN BLOCK

*A child finds a block with a chipped corner. She is concerned that someone might get hurt on the rough spot. How can this be prevented?*

| Children's Possible Solutions | Your Role |
|---|---|
| 1. A small group wants to take the block to the carpentry bench and sand off any sharp splinters. | Support the children's interest in safety. Provide goggles and sandpaper. |
| 2. One child suggests checking other blocks for dents or chips. | Help the child organize a "search party." Provide unbreakable magnifying glasses to observe small trouble spots. |
| 3. Others think they need rules for handling the blocks: No throwing blocks into the cabinet at cleanup time. | Brainstorm ideas about properly caring for equipment. |

## Extending the Experience

- Set up basins of soapy water. Use brushes dipped in bubbles to clean the grime off wooden blocks.
- Set up a repair center for other broken items from the room or outdoors.
- Ask your children to dictate a letter to their families about taking home broken toys to fix.

# QUICK CLEANUP

*Everybody is dawdling. Nobody is picking up the blocks. It's time to go home. How can the mess get cleaned up quickly?*

| Children's Possible Solutions | Your Role |
| --- | --- |
| 1. "Let's make it fun with riddles," announces one child. Then he says, "Can you find the triangle ones?" | Encourage everyone to think up other classifications (size, color, etc.) for their riddles. |
| 2. "We can fill up our toy vehicles with blocks and drive them over to the shelves," proposes another child. | Support the children with your physical presence. Verbally cheer them on. |
| 3. Others suggest having a race to see if the blocks can get put away before the cleanup song is over. | Help the race to go faster by showing the children how to form small groups to create assembly lines. |

## Extending the Experience

- Cut different block shapes out of construction paper. Mount the shapes on the block shelves to indicate where the various blocks belong.
- Take a field trip to a construction site to see how the workers pick up their materials (dump truck, wheelbarrow, crane). Make a class mural showing the different ways.
- Discuss some ways to make block cleanup easier, such as, "Only get out as many blocks as you need," or, "Everyone puts away five blocks."

# CAVES AND TUNNELS

*Three children wish to create individual secret hideaways. However, they want to be able to communicate with each other. How can they do this?*

| Children's Possible Solutions | Your Role |
|---|---|
| **1.** Two children suggest building caves by using the large hollow blocks for the cave walls and blankets for the roofs. | Compliment the children on their clever idea of combining the two materials. Acknowledge the children's attention to safety by using a "soft" roof. |
| **2.** Another child says they could use the large cardboard blocks to connect the caves with tunnels. | Help the child clarify the size of the tunnels—are they for talking through or crawling through? |
| **3.** All three think the caves might be too dark. They want to use flashlights. | Provide the requested props. Brainstorm other relevant items (sleeping bags, maps). |

## Extending the Experience

- Use shoeboxes, clay figures, and magazine pictures to create dioramas. Cut small peepholes in the covered diorama boxes to view the "secrets" inside.
- Work in small groups to make secret tunnels and caves at the sand table. Provide cylinders, small table blocks, and other materials as needed.

**Skills Used**
Social-Emotional Skills
Creative Thinking Skills

# SAVE OUR HOSPITAL

*The children in the morning group worked long and hard to create a very large, comprehensive hospital structure. Even though they know the afternoon class needs the blocks, they do not want to take down this project of which they are so very proud. What can they do?*

| Children's Possible Solutions | Your Role |
| --- | --- |
| 1. Some children want special permission to leave the structure up. The group suggests making a sign: "Please come look at our great hospital, but don't take it down." | Agree to honor this particular request. Acknowledge the children's outstanding collaboration and persistence. Write their dictated words on a sign. Let them post the sign near their creation. |
| 2. Another child suggests taking instant photos of the medical facility so everyone can remember it. | Offer to take dictation about the project, then put the photos in an album to keep in your library. |
| 3. Some propose that at dismissal time, they could bring their parents on an "open house" hospital tour. | Expand on this idea. Suggest videotaping the tour to share later with parents who couldn't come to the open house. |

## Extending the Experience

- Create "Come See Our Hospital" posters or invitations with beautiful paper and brand new crayons.
- Offer props to encourage hospital-theme dramatic play (medical uniforms, toy ambulance, gauze masks, etc.).
- Invite medical personnel (X-ray technician, nurse, pharmacist) to visit and talk about their jobs or health-related issues.

**Skills Used**
Social-Emotional Skills
Creative Thinking Skills

# GIRLS STAY OUT!

*The boys are planning to build a block fort together. The girls want to join in. The boys tell them they can't. They say, "You have to be strong. This is 'boy' stuff!" How can this conflict be resolved?*

| Children's Possible Solutions | Your Role |
|---|---|
| 1. One boy thinks maybe the girls could do something that didn't require great strength—like being cooks for the fort. | Discuss with the boys how important it is to look at a situation from all viewpoints. How might the girls feel? Explain how both men *and* women can do all sorts of jobs. |
| 2. A girl suggests that they could build the town that surrounds the fort. | Compliment the girl on her ability to find a solution. Talk about it with everyone. |
| 3. Several girls say that if everyone works together, boys *and* girls, the fort can be built even bigger and faster. | Encourage the children to pursue this idea. |

## Extending the Experience

- Provide lots of open-ended activities and materials (sand, water, clay) that boys and girls can use successfully as they work together.
- Ask parents to share their careers with the children. For example, an artist could set up an easel and paint alongside the children.
- Place posters and pictures in various classroom centers depicting girls and women and boys and men in less traditional roles (a female carpenter or a daddy diapering a baby).

# NO MORE BLOCKS

*While building "Dinosaur Land," the children find there are no more blocks on the shelves. They look around and see a girl hoarding a small pile of blocks in front of her. They want those blocks! What can they do?*

| Children's Possible Solutions | Your Role |
| --- | --- |
| **1.** Because they have lots of builders and they can't finish their project, the dinosaur group thinks the girl should share her blocks with them. | Help both sides look at and discuss the problem from each other's point of view. Is the girl planning to use the blocks? Does she want to share? Could they include her? Might they trade some things? |
| **2.** Another child begins negotiating with a friend to see if they can take some sections of Dinosaur Land apart and use the blocks elsewhere. | Help the children keep their negotiations focused. |
| **3.** Others bring boxes from the home center. They discuss using these like blocks. | Encourage the children's creative thinking with open-ended questions: "How could you use the small box?" |

## Extending the Experience

- Let your children make their own blocks out of soft wood scraps and sandpaper at the carpentry bench. Talk about shapes, sizes, and textures of the blocks.
- Set out dinosaur models in the sand and water areas outside.

**Skills Used**
Social-Emotional Skills
Creative Thinking Skills

# CRASH!

*Each time a block tower is built, someone crashes it down. The builders are frustrated and angry! What can they do?*

| Children's Possible Solutions | Your Role |
| --- | --- |
| **1.** One child says he loves the loud crashing noise and knocks the blocks down on purpose. The others say he is a "dummy" and shouldn't be allowed in the block center. | Calm everyone down. Raise the children's awareness of each other's views. Discuss their feelings about name calling and destroying projects. Designate an area where children may safely crash their own work. |
| **2.** Others say they want some rules, such as: Ask permission to crash other's blocks. Only crash your own blocks. | Discuss natural and logical consequences for their rules. |

## Extending the Experience

- Set out newspaper, towels, and bubble wrap. Experiment to see which ones muffle the sound of crashing blocks best.
- Create spontaneous, musical, crashing sounds with cymbals. Try out other "noisy" instruments.

# PARKING GARAGE

*The garage the children built isn't large enough to hold all of their cars. One child explains, "We need a parking garage." The others agree, and together they build a two-story garage. Now, how can they get their cars up to the second level?*

| Children's Possible Solutions | Your Role |
|---|---|
| **1.** One child suggests an elevator, but says he doesn't know how to make one. | Offer pulleys and some string for the children to experiment with. How could they raise the cars with these? |
| **2.** A girl describes how her father makes his car go higher on a ramp when he changes the oil. | Encourage the children to see if they can find any blocks that would work like ramps. Introduce new vocabulary words such as *force, gravity, inclined plane,* and *motion.* |
| **3.** Another child says the place where his mom parks her car has a road that goes around and around. | Help the child clarify his idea. Does the road merely go around, or does it slant upwards as well? What blocks might work? |

## Extending the Experience

- Design a child-size obstacle course for your children using all of the big blocks. Be sure to include various ways of going up and down.
- Find items in the room to use as ramps. Race miniature cars down the ramps.
- Use cardboard boxes to make garages with movable doors. Decorate the garages with paint.

# WHICH IS LONGER?

*One group of children has constructed a set of railroad tracks on one side of the block area. Another group has designed a superhighway going in another direction. They are having an argument over which road is longer. How can they decide?*

| Children's Possible Solution | Your Role |
| --- | --- |
| **1.** "Let's count the blocks in each," says one child. | Ask open-ended questions to help the children focus on the blocks: "Why do you think it makes a difference whether all the blocks used are the same size? What happens if they are different sizes?" |
| **2.** A girl suggests that she and some friends could lie down on the floor beside the railroad tracks and then beside the road to see which one uses more people. | Encourage trying out this innovative, collaborative effort. |
| **3.** Several children indicate they want to use something to measure the track and the road so they can compare the lengths. | Expand on this idea. Provide unconventional measurers, such as a length of string or a shoe, and standard measurers, such as a yardstick, a measuring tape, or a ruler. |

## Extending the Experience

- Measure and compare distances on the playground. How many steps does it take to reach the fence or how many children to hug a tree?
- Measure the children's heights and graph them.

# THE INTERSECTION

*Two groups are using blocks to construct roads from opposite sides of the room. Several builders begin to wonder how they will get their roads to join together. What can they do?*

| Children's Possible Solutions | Your Role |
|---|---|
| 1. One child worries that there will be some space left between the roads. He thinks they should use small blocks to finish the project. | Compliment the child on his ability to think ahead and visualize the completed roads. Provide the necessary blocks. |
| 2. A few children suggest making an interchange like the ones they've seen on the big highways. They get several curvy blocks to use as connectors. | Comment on the children's flexibility. |
| 3. Another child wants to move one of the roads over with a toy bulldozer or a crane—like real road construction workers do. | Share the child's excitement in applying a workable idea that she has observed elsewhere. |

## Extending the Experience

- Strengthen hand-eye coordination by creating patterns with parquetry blocks.
- Create colorful mosaics with small, colored plastic tiles or small construction paper squares.
- Practice spatial relationship skills with an assortment of puzzles.

# HINGES AND DOORS

*The children build a large house with hollow blocks. How can they make a door for their house that will open and close easily?*

| Children's Possible Solutions | Your Role |
| --- | --- |
| **1.** Someone suggests using a cardboard rectangle cut out of a large cardboard box. | Be available to assist if the children need the thick cardboard cut to a specific size. |
| **2.** One child proposes using tape to "hinge" a sheet of posterboard to the blocks. | Provide masking tape and posterboard. Compliment the child on his ingenuity. Discuss the word *hinge*. |
| **3.** A girl shows how to place a towel over the opening and roll it up and down like a shade. | Facilitate with a directive type of question: How is it different when you use the towel than when you use the cardboard? |

## Extending the Experience

- Set up a hinge display in the science discovery area.
- Supply small pieces of cardboard and masking tape in the art center.
- Look at a book of different types of houses—wood frame, cave, tepee, yurt, etc.—and discuss their different types of doors.

# BOATS IN THE MARINA

*One child is very excited about her recent sailboat trip. She wants her friends to help her create a marina. Unfortunately, there aren't any boats available in the block center. What can they use instead?*

| Children's Possible Solutions | Your Role |
|---|---|
| **1.** Some children want to make sailboats by decorating unit blocks with construction paper shapes taped to straws. | Enjoy the children's enthusiasm! Provide tape, scissors, paper, and collage materials to add even more color. |
| **2.** A boy thinks they should use blue crepe paper or blue cellophane paper for the water. | Show the child the various materials available in the art closet. Discuss the merits of each substance with him. |
| **3.** Some say they aren't sure what a marina looks like. Should they add bridges and places to tie up boats? | Provide reference materials, such as posters of waterfronts and books with pictures of docks, lighthouses, and ships. |

## Extending the Experience

- Outdoors, place a water tub in the sandbox for floating small boats. Create a miniature harbor city in the sand with table blocks.
- Make unique sailboats with plastic-foam food trays, stickers, and colored tissue paper.
- Play a game in the dark with a flashlight "lighthouse." Safely guide the "ships" into the rocky harbor.

# THE WINDOWS ON THE BUS

*The children think it would be fun to create a bus with blocks. How can they make windows to see out of?*

| Children's Possible Solutions | Your Role |
|---|---|
| 1. One group uses unit blocks to build a solid wall for the side of the bus. As they pull out one of the unit blocks to create the window opening, the wall falls down. | Help the children make the connection between the alignment of the blocks and the collapsed wall. Tell them to keep trying. |
| 2. Another group watches as the wall falls down. They try building a wall and leaving open spaces for the windows as they go along. | Compliment the children on their ability to watch and learn. |
| 3. A girl says they should tape plastic food wrap like her dad uses in the kitchen over the openings to make them look like real windows. | Provide the necessary materials. Acknowledge her creative thinking. |

## Extending the Experience

- For dramatic play, add passenger seats to the bus. Provide other props, such as a bus driver's cap and a fare box.
- Draw maps to follow on their pretend bus trips.
- Sing "The Wheels on the Bus" and make up new verses.

# TEDDY BEARS' HOUSE

*The children wish to create an elegant house for their beloved teddy bears. How can they add special features, such as stairs and beautiful furniture?*

| Children's Possible Solutions | Your Role |
|---|---|
| **1.** Some children try to create stairs by stacking up several blocks, and then pushing each one back a little. Unfortunately, the blocks all fall down. | Let the children know it is important to experiment. Take them on a walk around the school to observe how other staircases have been built. |
| **2.** A few other children think it would be fun to create tables and beds with the wooden unit blocks, then decorate them with small, colorful blocks. | Applaud the children's inventiveness in combining different types of blocks. |
| **3.** In the art center, other children find wallpaper samples to beautify the walls, and lacy fabric samples to make lovely, airy tablecloths and curtains. | Encourage the children's whimsical thinking. Sit on the floor and play, too. |

## Extending the Experience

- Set up a tea party for teddy bears. Dress up the bears and borrow cups and saucers from the dramatic play center.
- Encourage your children to act out the parts as you tell the story of *Goldilocks and the Three Bears.*
- Cut out pictures of home furnishings from catalogs and magazines. Tape the pictures to blocks to create fashionable accessories for the block center.

# THE COUNTY FAIR

*The children find a basket of ribbons, rickrack, sequin strands, braid, and beads. They want to use these materials to make a county fair. How can they do this?*

| Children's Possible Solutions | Your Role |
| --- | --- |
| **1.** A few children build animal barns and judging arenas for toy farm animals. They use the braid for halters and tie on ribbons for the awards. | Encourage and discuss the children's creative manipulations. |
| **2.** For excitement on the midway, other children build a merry-go-round, drape it with sequin-strand "lights," and add toy zoo animals for riding on. | Provide some calliope music for the carousel to heighten the effect. |
| **3.** Some children make amusement stands and decorate them with rickrack. Toy people play the games and are awarded bead necklaces when they win. | Join in the fun at this lively, imaginative county fair. |

## Extending the Experience

- Bake (and sample!) cookies at the food judging booth.
- In the writing and art centers, create tickets and posters for the fair.
- Test skills by trying to knock down plastic bottles with a rubber ball.

Problem Solving in the

# SCIENCE AREA

# MY HANDS ARE FREEZING

*The children come in from outside where they have been playing in the cold winter weather. Their hands are very cold. How can they warm their hands?*

| Children's Possible Solutions | Your Role |
|---|---|
| 1. One girl has seen her dad rubbing his hands together when they are cold. She thinks everyone should do this. | Encourage everyone to try this. Does it work? Discuss why or why not. |
| 2. A boy says they should all put their mittens back on. | Support this action. Discuss how mittens keep hands warm. Does the warmth come from the mittens? |
| 3. One child suggests washing their hands in warm water. | Explain to your children the danger of putting cold hands in warm water. Instead, let them wash their hands in cool, not cold, water. This water will feel warm to their very cold hands. |

## Extending the Experience

- Make a mitten matching game with your children's mittens.
- Collect mittens to donate to a local shelter.
- Have your children sit in pairs. Let them take turns massaging their partner's hands.

# THE HEAVY SPONGE

*While playing at the water table with sponges that are all the same size and shape, your children think they have observed that a wet sponge is heavier than a dry sponge. How can they find out for sure?*

| Children's Possible Solutions | Your Role |
|---|---|
| **1.** Your children feel the weight of dry sponges before dropping them into the water. Then they remove the sponges and feel their weight again. | Provide many sponges. Encourage the children to talk about their work. |
| **2.** One child brings the balance scale to the water table area. She compares wet and dry sponges on the scale. | Acknowledge the child's resourcefulness for using the scale. Discuss her findings. |
| **3.** Other children fill some sponges with water and weigh them. They squeeze out the water and weigh them again. | Talk about the children's findings and record them on a chart. |

## Extending the Experience

- Provide a variety of sponges for wiping up spills and washing classroom items.
- Weigh containers, some empty and some full of water. Talk about the differences in weights.

# FEEL THE BEAT

*Your children have been moving to music—pretending to be jumping beans. One child puts his hand on his chest and feels a strong, rapid thump-thumping. He announces this to his classmates, and all the children put their hands on their chests. They want to know what is causing this. What can they do?*

| Children's Possible Solutions | Your Role |
| --- | --- |
| **1.** A girl suggests that maybe her thumping is being caused by her breathing. She holds her breath to make it stop. | Encourage the girl to experiment. When she holds her breath, does the thumping stop? |
| **2.** After a minute, the boy says he can no longer feel the thump-thumping. Some children say sitting still will make it stop. | Foster discussion. Do the children feel a thumping in their chest when they are sitting quietly? |
| **3.** Another boy begins jumping up and down. "It's back!" he announces, a hand on his chest. | Praise the child for his quick thinking. Why does he think the thumping came back? |

## Extending the Experience

- Listen to heartbeats with a stethoscope.
- Beat out sounds like heartbeats on drums made from coffee cans or oatmeal boxes.

# THE SCRATCHED FLOOR

*Some children discover that the floor is very scratched in the area around the sand table. They want to know what scratched up the floor. How can they do this?*

| Children's Possible Solutions | Your Role |
| --- | --- |
| **1.** One child thinks the legs of the table scratched the floor. She looks for "scratchy things" at the bottom of the table legs. | Discuss whether sand table legs might make scratches when other table legs have not done so. If so, why? |
| **2.** The children decide to explore the floor around the rest of the room. They look around and don't find the same scratches. They think maybe the sand caused the scratches on the floor. | Comment on the children's investigative abilities. |

## Extending the Experience

- Offer plastic magnifying glasses for examining grains of sand.
- Provide various grades of sandpaper for sanding blocks of wood or other materials.
- At the sand table, draw "scratches" in the sand with craft sticks.

# A HAIR-RAISING EXPERIENCE

*On a cold, low-humidity day, several children are sliding down the hard plastic sliding board. Another child walks under the board as they are sliding and feels his hair being pulled upward. He calls for the other children to come and see. They want to know what made his hair stand up like that. How can they do this?*

| Children's Possible Solutions | Your Role |
|---|---|
| **1.** So many children are now standing under the slide, no one is sliding down. The children notice that the boy's hair is no longer standing on end. They think sliding makes his hair stick up. | Discuss this change. What might have caused it? |
| **2.** Other children think that it is the play equipment that makes his hair stand up. They ask him to stand under the climbing bars to see what will happen. | Encourage the children's investigation. Talk about the results. |

## Extending the Experience

- Investigate static electricity. Run a plastic comb through your hair several times. Move the comb across several small pieces of tissue paper or facial tissue. Watch the paper pieces "jump" up to the comb.
- Make "static hair" pictures. Glue magazine-picture faces on construction paper and add short pieces of yarn for "stand up" hair.

# DIG THIS!

*Your children are digging in dirt. In her shovel, one child sees a small worm. "What's this doing in the dirt?" she asks. The children are curious. Do some creatures live in the dirt? How can they find out?*

| Children's Possible Solutions | Your Role |
|---|---|
| **1.** The girl puts the worm on the dirt to see if it will go into the ground or crawl to some other place. | Ask the girl why she thinks the worm crawled where it did. |
| **2.** A boy gets a magnifying glass to look for other creatures in the dirt. | Discuss why the boy found these animals underground. Why would the ground be a good, safe home for them? |

## Extending the Experience

- Move along the floor like worms.
- Purchase an ant farm for your room. Have your children discuss what they observe.
- Read about worm composting bins. Consider adding one to your outdoor classroom area.

# THE BEAR'S CAVE

*After reading a book about a bear and its cave, some children decide they want to make a cave. How can they do this?*

| Children's Possible Solutions | Your Role |
| --- | --- |
| 1. Some children go under a table, but decide it is not a cave because it has no walls. They bring blankets from the dramatic play center and drape them over the table, providing sides. | Admire the children's efforts. |
| 2. Another child says it is not dark like a cave. He turns out the classroom lights. This results in howls from children in other centers. They want the lights on so they can work in their centers. | Help the children negotiate a solution. Is there a way to darken the cave without darkening the entire room? |
| 3. A girl who has seen a movie about caves says the walls in a cave are black. She thinks that by making black walls, they can create a cave. | Support the girl's efforts to find dark materials for the walls of the cave. |

## Extending the Experience

- Display pictures of real caves and caverns.
- Hide teddy bears around the room and let your children go on a bear hunt. Provide enough bears so that each child can find one.
- Go on a field trip to a zoo and watch real bears, if possible. Where do the bears sleep?

**Skills Used**
Social-Emotional Skills
Motor Skills
Critical Thinking Skills

# SQUIRT BOTTLES

*Your children are using turkey basters and squeeze bottles at the water table. A boy begins squirting water at the others, wetting their clothes. What can the other children do?*

| Children's Possible Solutions | Your Role |
|---|---|
| 1. Several children giggle and squirt water back at him and at each other. | Stop the water action. Remind the children that the water must stay in the table. Brainstorm with the children to think of ways they could squirt water and keep it in the water table. |
| 2. One girl starts crying. She thinks children who are squirting water should be removed from the water table. | Discuss this with the children. What do they think of this idea? |
| 3. Another child says we need a rule: No squirting. | Talk about this possibility. Are there acceptable ways to squirt the water? |

## Extending the Experience

- Put thinned tempera paint in squirt bottles. Let your children make designs by carefully squirting the paint onto large sheets of paper. Remind them to keep the paint on the paper.
- Provide water and squirt bottles for outdoor play.
- Set out eyedroppers and small containers of colored water for dripping colors into the water in the water table.

# DISAPPEARING SHADOW

*Your children are playing a shadow game outside: chasing and stomping on each other's shadows. A girl steps into the shade of a tree and her shadow disappears. She wants to know where it went. What can she do?*

| Children's Possible Solutions | Your Role |
| --- | --- |
| **1.** The girl asks her friends to come help. Some of their shadows disappear, but children standing in the direct sunlight shout with glee, "We still have our shadows!" | Encourage discussion. What do the children think happened? |
| **2.** The girl runs out of the tree's shade. She sees her shadow emerge from the tree's shadow. She moves in and out of the shade, making her shadow appear and disappear. | Comment on the girl's observation skills. |

## Extending the Experience

- Help your children create outdoor games using their shadows.
- Place butcher paper on the floor. Have a child stand so his shadow falls on the paper. Trace around the shadow.
- Using an overhead projector, project shadows onto a wall. Ask your children to guess what object is making each shadow.

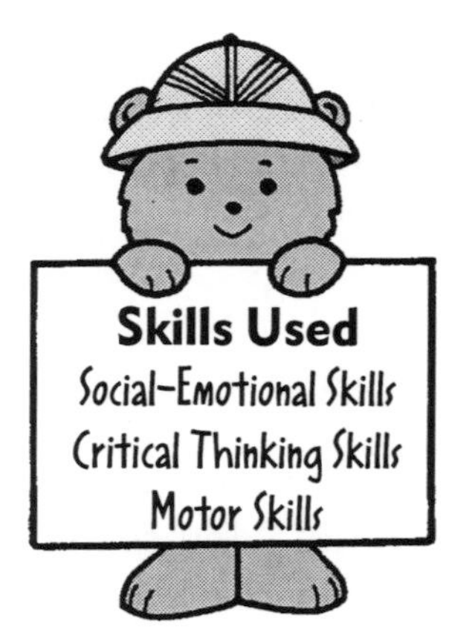

# UNPOPPED CORN

*Your children have been watching popcorn popping in a hot air popper. When the popping is finished, they discover some unpopped kernels left in the machine. How can the children find out why this happened?*

| Children's Possible Solutions | Your Role |
| --- | --- |
| **1.** Some children think the machine was turned off too soon. They ask the teacher to plug in the machine so that the remaining kernels have a chance to pop. | Remind the children that the corn had stopped popping when you unplugged the machine. Explain that it is not safe to try to force unpopped kernels to pop. Discuss what might have caused these kernels to remain unpopped. |
| **2.** Another child thinks this batch of popcorn had a problem. She suggests popping another batch so that all the kernels can pop. | Pop another batch of popcorn. Discuss the results. |
| **3.** Several children say that these kernels are different. They want to look in the bag of popcorn kernels to see if some of the kernels look different from the others. | Have the children take out the "different-looking" kernels. Put these in the popper. What happens? |

## Extending the Experience

- Discuss how heat changed the popcorn kernels. Discuss how other foods are changed by heat, such as cheese on toast, apples cooking into applesauce, or a hard potato softening when it is baked. Try one of these foods for a snack.
- Fill clear plastic cups with potting soil and plant some popcorn kernels. Add water and sunshine.

# THE WIND

*Your children are playing outside on a windy day when some paper is invisibly picked up and scattered across the yard. One child notices that when the papers fly to one side of the building, they stop moving. The children want to know why. What can they do?*

| Children's Possible Solutions | Your Role |
|---|---|
| **1.** Someone shouts, "An invisible monster moved the papers!" The other children run around searching for the invisible monster. | When the children's search turns up nothing, encourage them to consider other explanations. |
| **2.** One child notices that when she stands in certain places, she can no longer feel the wind. Other children join in and begin testing how it feels when they stand by trees, under the swings, by the climbing bars, etc. | Discuss the children's findings. Invite them to discover why they feel no wind in those places. |

## Extending the Experience

- Make simple paper-bag kites. Fly the kites while running *with* the wind and running *into* the wind.
- Have your children move to music as they pretend to be trees swaying or leaves scattering in the wind.
- Hang a flag or a banner outside and watch it flap in the wind.

Skills Used
Critical Thinking Skills
Motor Skills
Creative Thinking Skills

# SINKING AND FLOATING BOTTLES

*Your children are playing at the water table. They notice several plastic bottles that look the same but act in different ways. Some are floating on top of the water, a few are lying at the bottom, and others are floating halfway in between the top and the bottom of the water. They want to discover why the bottles are acting differently. How can they do this?*

| Children's Possible Solutions | Your Role |
|---|---|
| **1.** A boy thinks that the bottles on top just haven't been pushed down into the water. He pushes one of the floaters to the bottom. It floats back to the surface. | Compliment the boy on his idea. Talk about the results. |
| **2.** Other children think the water is making the bottles sink. They fill all of the floaters with water and replace the lids. They put them in the water table and watch them sink to the bottom. | Discuss the children's actions and observations. What can they conclude? |

## Extending the Experience

- Set out more unbreakable containers with lids for additional water table experiments.
- Add ready-made sieves and colanders to the water table, or make your own by punching holes in the bottom of plastic containers.
- Provide aquarium gravel for additional sinking and floating experiments at the water table.

# Grass Stains

*As your children are running, one child drops to his knees and slides across the grass. When he stands up he notices that the knees of his tan pants are now green. How can he find out what happened?*

| Children's Possible Solutions | Your Role |
|---|---|
| **1.** The boy picks up some grass in his hand and rubs it on his pants. | Compliment the boy on his inquisitiveness and suggest he rub the grass on a towel or rag instead of his pants. |
| **2.** The boy takes a white rag and rubs it vigorously on a patch of grass, then looks at the rag. | Talk about what happens to the rag. Where did the color come from? |
| **3.** Taking a sheet of white paper from the art center, the boy places the paper on the grass. He stands on the paper and twists his feet back and forth. When he looks at the paper, he discovers green stains on it. | Provide materials and discuss the results. |

## Extending the Experience

- Provide a tub of soapy water so your children can take turns trying to wash the stained rag. Is it easy to remove the green stain?
- Glue grass clippings to heavy paper to make green designs.
- Experiment with making stains on paper with other nature items, such as dandelion flowers and blackberries. (Caution: These stains tend to be permanent.)

Motor Skills
Critical Thinking Skills
Creative Thinking Skills

# CHANGING CONES

*Your children have been playing outside with pine cones for several days. After a big rainstorm, they discover that the pine cones look different. The cones have tightly closed up their scales. The children want to know why this happened. What can they do?*

| Children's Possible Solutions | Your Role |
| --- | --- |
| **1.** One child finds some dry pine cones, still open, inside the storage shed. She suggests taking them inside and putting them in the water table. | Support the child's investigation. Discuss the results. |
| **2.** Another child thinks that if they bring the wet pine cones inside to dry, the scales will open again. | Provide a dry place to set the pine cones. Talk about what happens as they dry. |

## Extending the Experience

- Shake a pine cone over a sheet of paper. What do your children observe? (Seeds fall out.) Let them plant the seeds.
- Collect cones from a variety of trees to compare and contrast.
- Make bird feeders out of pine cones by spreading on peanut butter and rolling them in birdseed. Hang them up outdoors.

# FADED PAPER

*Your children find some construction paper that has been lying on the windowsill. Part of the paper was covered by a box. They notice that the uncovered portion of the paper is lighter in color. They want to know how this happened. What can they do?*

| Children's Possible Solutions | Your Role |
| --- | --- |
| **1.** A boy thinks someone erased part of the paper. He gets a new sheet of construction paper and begins erasing some of it. | Encourage the child's comparison. Talk about his findings. |
| **2.** Some other children think the sun faded the paper. They put a new sheet of construction paper on the windowsill and cover part of the paper with wooden blocks. | Compliment the children on their creative investigative techniques. Ask how often they will check their papers for changes. Discuss the changes as they discover them. |

## Extending the Experience

- Make sun-print art by placing sheets of construction paper in a sunny place and arranging interesting objects on them. In a few days, remove the items and enjoy the sun-print pictures.
- Investigate things that are always in the sun (a building, a road, a fence, etc.). Look for signs of fading.

**Skills Used**
Creative Thinking Skills
Critical Thinking Skills

# HANDPRINTS

*Your children are playing at the sand table. One child presses her hand into damp sand and sees that she has made a handprint. She and her friends make more prints in the sand. They wonder if they can make handprints in other places. How can they find out?*

| Children's Possible Solutions | Your Role |
| --- | --- |
| 1. The children press their hands into the loose dirt in the vegetable garden to see if they leave prints there. | Compliment the children on their ingenuity. |
| 2. At the water table, the children press their hands into the water. No handprints remain. | Encourage the children's inquiry. Discuss the results. |
| 3. One child presses his hand into a dishpan of plastic foam peanuts. When he removes his hand, there is no handprint. | Suggest that the child try other materials. |

## Extending the Experience

- Press your children's hands into wet plaster of Paris. Label each handprint. (When using plaster of Paris, follow the directions on the box for preparing it and for disposing of any leftover plaster.)
- Press hands into shallow pans of paint, then onto sheets of paper to make handprints.
- Outside, provide water and let the children discover that their wet hands leave temporary handprints on some surfaces (cement walkways) but not on others (grass).

# BERRY BUSHES

*Outside, your children discover a bush full of berries. What can they do with them?*

| Children's Possible Solutions | Your Role |
|---|---|
| **1.** Several children suggest picking the berries to eat. | Caution all of your children to *never* eat growing things without checking first with a grownup to see if they are safe. Allow them to pick and eat these berries only if you know for certain they are edible. |
| **2.** One girl stomps some of the berries that have fallen to the ground and is fascinated with the stains they make. | Offer a plastic magnifying glass so the child can investigate close up. Discuss her findings. |
| **3.** A pair of children discover seeds inside the berries. They want to plant them to see what grows. | Provide materials for planting and observing growth. |

## Extending the Experience

- Offer several varieties of berries for snack. Compare taste, texture, size, and color.
- Let your children help you make blueberry pancakes.
- Observe birds that are interested in the berry bush. What are they doing? Talk about outdoor plants that provide food for animals.

# RAMPS AND CARS

*Several children are playing with large hollow blocks, boards, and small cars and trucks. Two of them lean boards at an incline against blocks, making ramps for their cars. The cars go down the ramp and partway across the floor. One travels farther than the other. How can they find out why?*

| Children's Possible Solutions | Your Role |
|---|---|
| **1.** The two children decide to build identical ramps and roll identical cars down them. The cars travel similar distances. | Discuss the results. Why did the cars travel the same distance? |
| **2.** One child builds a tall ramp. Another child builds a short ramp. They roll identical cars down the ramps at the same time. One car travels a long distance, the other a short distance. | Encourage discussion of the benefits of long and short ramps. Introduce the concept of incline and suggest they test big inclines and little inclines. |

## Extending the Experience

- Outdoors, set out large pieces of cardboard and several cars and trucks. What games can your children invent?
- Try other items on the ramps. Use items that roll and others that do not. How do the results differ?
- Provide big and small items to roll down identical ramps. Which ones roll the fastest? Which ones go the farthest?

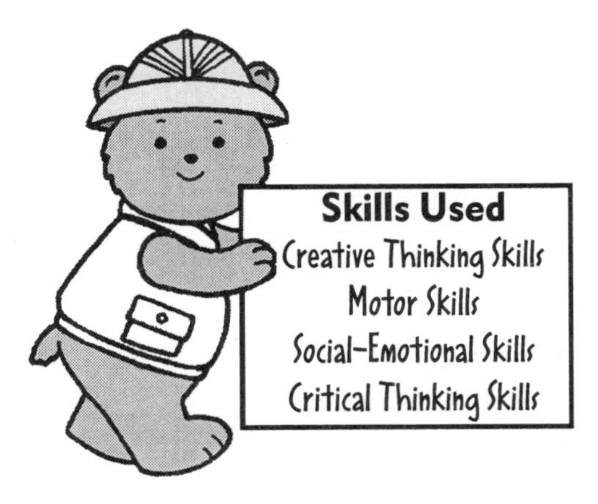

Problem Solving

# OUTDOORS

# A FLAT BALL

*Your children discover that their big rubber ball is flat. It won't bounce. How can they play with it?*

| Children's Possible Solutions | Your Role |
| --- | --- |
| **1.** After holding the flat ball for a while, your children realize that they can grasp it by the wrinkles. They play catch by tossing the ball up into the air. One child suggests throwing it sideways like a flying disk. | Comment on your children's creative suggestions for turning a broken toy into a useful one again. |
| **2.** While sitting in a small circle, several children make up a silly game by passing the flat ball around like a "hot potato." | Join the fun as a player! |
| **3.** A girl shares this idea: When her dad's tire goes flat, he fills it up with air at the gas station. The group decides to take a trip to a gas station to put air in their ball. | Make and carry out arrangements for visiting the gas station. Check to see that you have written permission from parents for the trip. |

## Extending the Experience

- Create balls with other materials, such as newspaper or aluminum foil.
- Find other objects to roll, such as pine cones or hoops.
- Form ball shapes with wet sand.

# WIGGLY SWINGS

*Your children notice that the swings frequently go in all directions, sometimes even bumping against other swings. Although some of the children think that this is funny, the seemingly uncontrolled movement frightens other children. How can they make swinging fun and safe for everyone?*

| Children's Possible Solutions | Your Role |
| --- | --- |
| **1.** Your children take turns sitting in different places on the seat of a swing while they pump their feet. Observers call out such comments as "The swing goes wiggly!" when a child sits too far to the right or left, or "The swing goes straight!" when a child sits in the middle. | Before each trial, encourage your children to guess what might happen. Help them discuss the results after each trial. |
| **2.** The children decide to see what might happen if the pusher tries moves such as these: stands directly behind the swing while pushing; stands to one side while pushing; pushes with just one hand; stops pushing. | Assist your children as they try out their ideas. Help them develop simple safety rules for swinging such as: Sit properly on the swing seat. Face the same direction as the other swingers. Allow only one person at a time on a swing. Stand directly behind a swing and use both hands when pushing it. |

## Extending the Experience

- See if the children's solutions will work with other types of swings, such as a tire swing or a rope swing.
- Using an empty tissue box and yarn, make a safe swing for a doll or a stuffed animal.

# THE TUNNEL COLLAPSES

*When your children try to make a tunnel by digging with their hands in dry sand, the walls of the hole quickly cave in. How can the children create an open tunnel for their toy cars?*

| Children's Possible Solutions | Your Role |
|---|---|
| **1.** Several of your children try to dig deeper and at different angles. | Acknowledge the children's persistence. |
| **2.** Two children look in the basket of sand toys to find something to hold the sand up. They discover a cylindrical potato chip container that is large enough for their toy cars to pass through. | Foster independence by providing easy access to various kinds of sand toys and other appropriate materials. Praise your children for making such a creative connection between a potato chip container and a tunnel. |
| **3.** Other children decide to add some water to the sand. Using hands and shovels, they try to pack the damp sand and dig a tunnel through it. | Encourage and discuss what the children do as they work. |

## Extending the Experience

- Sing and act out the song "London Bridge."
- Use blocks and other manipulatives indoors to create tunnels for toy cars.
- Set out large, open-ended boxes or commercial, flexible, cloth cylinders to use as tunnels to crawl through.

# THE WAGON IS STUCK

*A girl tries to pull two other children in a wagon over a pea gravel path. The wagon gets stuck. What can they do?*

| Children's Possible Solutions | Your Role |
|---|---|
| **1.** The girl tells the other children to get out. She thinks that they will need to fix a flat tire. | Encourage the children to check the wagon, then acknowledge the girl's logical thinking. |
| **2.** One child says that there are too many people in the wagon. He thinks one person should get out and push while the girl is pulling. | Help your children make connections between lightening the load and adding more power. |
| **3.** A rider remembers that the wagon moved all right on the blacktop. She suggests taking the wagon back there. | Challenge your children to test the wagon on various surfaces, then discuss their observations. |

## Extending the Experience

- Take stuffed animals for an adventurous wagon ride.
- Make up verses and sing "The Wheels on the Wagon" to the tune of "The Wheels on the Bus."
- Fill the wagon with different materials such as blocks or buckets of sand. Each time, see how easy or hard it is to pull the wagon.

# YOU CAN'T CATCH ME!

*When your children first go outside, a number of them immediately engage in chasing games. Often the running escalates to a roughhouse type of play, which excites some children but intimidates others. What can they do to make playtime enjoyable for everyone?*

| Children's Possible Solutions | Your Role |
| --- | --- |
| **1.** Some children suggest creating a "safety zone" to run to when they need to rest or don't want to play. | Help the children determine a place for the safety zone. Provide yarn or rope for outlining the area in case someone needs to quickly identify it. |
| **2.** One child suggests, "Yell 'Stop!' if you don't want to play." | Encourage the children to come up with other empowering words, such as "halt" or "freeze." |
| **3.** Other children discuss making a rule to allow touching only on the arms during chasing games. | Help the children practice gentle touches or quick taps on the arms. |

## Extending the Experience

- Play various types of tag, such as Hold Tag or Shadow Tag.
- Together, set up a jogging course on the playground.
- Try different ways of moving across the playground, such as galloping, hopping, or pretend skating.

**Skills Used**
Social-Emotional Skills
Motor Skills
Critical Thinking Skills

# HERE COMES THE BALL

*When several of your children play with the rubber balls, they kick them into the sandbox, knocking down constructions. Or, they bounce the balls in front of the trike riders, causing traffic problems. This makes the other children angry! What can they do?*

| Children's Possible Solutions | Your Role |
|---|---|
| **1.** Some children are very upset and want to keep all balls off the playground. | Facilitate an exchange of viewpoints. For instance, how might such a rule make the ball players feel? What rules might help everyone on the playground? |
| **2.** One child suggests making separate ball fields for different ball activities. | Challenge your children to find materials to use to create boundaries for the ball fields. For instance, they might use plastic cones for the ball-kicking area or fabric strips for the ball-bouncing area. |
| **3.** Another child says, "Kick the ball *away* from the sandbox." | Encourage your children to try out and evaluate how playing ball in different directions or locations affects others on the playground. |

## Extending the Experience

- Bowl, using rubber balls and plastic bottles.
- Bounce balls to music.
- Play catch with foam balls.
- Sit in a circle with legs outspread and quickly roll balls back and forth.

# KING OF THE JUNGLE GYM

*One child always bullies the others. He climbs to the top of the jungle gym, shakes his fist, and yells, "I'm the king! You can't come up!" What can the other children do?*

| Children's Possible Solutions | Your Role |
|---|---|
| 1. Several children say, "We should stay away. He's mean and might hurt us." | Help your children explore their feelings. Do they really want to stay away, or do they want to play with the equipment? |
| 2. Other children say that they should try to talk to the "king" or even yell back. | Demonstrate negotiation skills by listening to different viewpoints, picking a solution, trying it out, and then trying another solution if the first one doesn't work. |
| 3. Two children suggest trying to climb up the other side of the jungle gym, away from the "king." | Help your children develop an exciting collaborative play scenario that the "king" might want to join or be invited to participate in. |

## Extending the Experience

- Play cooperative games, such as Hokey-Pokey, in which no one dominates.
- Plan activities, such as Mirror Image, in which the children take turns being leaders and followers.
- Act out stories in which the bully is given a non-threatening role, such as that of a fluffy kitten.

# MY GOLD MINE

*Your children enjoy burying small toys in the sandbox and then digging for "gold." Lately, a girl has been hoarding the gold and making the other children upset by announcing, "This is my gold mine!" What can the other children do?*

| Children's Possible Solutions | Your Role |
|---|---|
| 1. One child says, "If she doesn't share, then she can't play with us." | Brainstorm with your children to see if they can find an agreeable compromise. |
| 2. Another child suggests setting time limits: "After one group plays with the gold, then another group takes a turn." | Help your children negotiate workable rules for turn-taking. |
| 3. Together, several children decide to use dandelions to make their own "gold." | Compliment the children on their flexibility and on their handling of the situation in a positive, cooperative way. |

## Extending the Experience

- Paint small stones or twigs gold to make "pieces of eight" to bury in the sandbox.
- Plan a treasure hunt. Draw symbol maps for the children to follow.
- Plan pair activities in which two children work together. For instance, during sand play, one child can hold a sieve while the other shovels sand into it.

# WE'RE THIRSTY!

*After playing outdoors for a while, your children often feel thirsty and ask one of the teachers to take them inside for a quick drink of water. This leads to minimal supervision on the playground. How can the problem be solved?*

| Children's Possible Solutions | Your Role |
|---|---|
| 1. One practical child says that everybody should remember to get a drink of water before going out. | Reinforce this idea by asking your children if they need a drink each time you prepare to go outdoors. |
| 2. Another child explains how his family always takes along a container of water when they go on outings. | Help your children apply the boy's idea. Supply a container, then ask for volunteers to help fill it with water and carry cups out to the playground. |
| 3. Several children suggest having snacks outdoors so that they won't have to go inside for drinks. | Provide a big cloth to sit on and fill a picnic basket with such items as juice, finger foods, cups, and napkins. |

## Extending the Experience

- Have a teddy bear tea party outdoors.
- During water play, use the water container to fill up various sizes and shapes of plastic containers.
- Hook up a hose to use to fill drinking containers with water.

# PUDDLES ON THE PLAYGROUND

*It has rained steadily for several days, and puddles are scattered over the hard surface of the playground. How can your children get rid of the water so that they can play without being splashed?*

| Children's Possible Solutions | Your Role |
| --- | --- |
| **1.** Several children say that they need a giant sponge or some towels or newspapers to absorb the water. | Encourage your children to collaborate and generate a list of absorbent materials. Provide requested materials, such as sponges, towels, or newspapers. |
| **2.** Another group suggests covering up the puddles by dumping dry sand on them. | Help the children open the sandbox and gather pails and shovels to use for carrying out their idea. Acknowledge their ingenuity. (Make sure that the wet sand is swept up as soon as possible to prevent slipping.) |
| **3.** A girl remembers seeing maintenance workers on TV sweep the water off a baseball field. She and a friend ask for brooms to use for sweeping away the puddles on the playground. | Provide the brooms. Discuss how exciting it is to use ideas seen elsewhere to create new solutions. |

## Extending the Experience

- Select a puddle and use chalk to trace around it each day. Observe how the puddle keeps shrinking, or evaporating.
- Use a variety of water toys, such as sponges, scoops, or basters, to remove water from a plastic tub placed on the ground.

# A SPLINTERY CLIMBER

*While descending the wooden climber, a boy gets a splinter in his finger. After you remove the sliver and cleanse the puncture, your children express concern about other possible splinters. What can they do to avoid them?*

| Children's Possible Solutions | Your Role |
| --- | --- |
| **1.** Some of your children suggest, "Stay away from the climber!" | Acknowledge the children's concern for safety. Ask open-ended questions to challenge their thinking such as "What else might you do?" |
| **2.** A girl notices that the injured child's finger is covered with an adhesive bandage. She suggests covering up the splinters on the climber by using tape. | Provide tape, such as transparent tape, masking tape, or duct tape, and safe scissors. Ask the children how long they think the tape will last if it should rain. Observe the tape after the next rainfall. |
| **3.** Other children want to highlight the splinter spots with paint to let climbers know about the "danger." | Provide bright, fluorescent paints, paintbrushes, and smocks. |

## Extending the Experience

- Identify other pieces of wooden equipment and carefully search for splinters.
- In the block area indoors, locate and sand rough spots on wooden blocks, trucks, and other toys.

# LOTS OF LEAVES

*When your children arrive on the playground, they find leaves everywhere, covering the ground, the sandbox, the bike path, and other play areas. What can they do with the leaves?*

| Children's Possible Solutions | Your Role |
|---|---|
| **1.** All the children agree to pick up the leaves and move them. Some suggest stuffing the leaves into bags; others want to rake them into piles. | Encourage your children to work together. Provide bags, rakes, and a wagon to haul the leaves away. |
| **2.** Several children demonstrate how much fun it is to jump into the leaf piles and cover themselves up. | Join in the fun! Help the children think of related action words. |
| **3.** A few children suggest recycling the leaves, like they do at home. | Help your children select a place for starting a compost pile. Encourage them to observe the pile throughout the year. |

## Extending the Experience

- Make leaf collections. Classify the leaves by shape, size, and color.
- Use paper and crayons to make leaf rubbings.
- Create leaf crowns to use for dramatic-play activities.

# MOVING THE ROCK

*A large rock is situated in the grassy play area. Your children want to move it so that they can create a ball field. What can they do?*

| Children's Possible Solutions | Your Role |
| --- | --- |
| **1.** Several children think that the rock is too big and too dangerous to move. They say that it should be left where it is. | Indicate how much you value the children's concerns about safety. Help them analyze ways to make the move safe. |
| **2.** Some of the children say that the rock is stuck in the ground. They want to dig around it. | Challenge the children to find different ways to dig. Provide tools they suggest, such as a shovel or a toy front-end loader. |
| **3.** Your children share what others have done. A girl reports that her dad pried out a rock. A boy says that his neighbor moved a rock by rolling it. | Introduce key words and encourage your children to select various equipment items. For instance, they could use a long board for a *lever,* or a wooden wedge for an *inclined plane*. |

## Extending the Experience

- Move pebbles or sand using various containers, such as coffee cans or cardboard boxes.
- Move items using wheeled toys, such as trikes or wagons.
- Use wind power to move bubbles or dandelion seeds.

# WE'RE SO HOT!

*Your children have been out on the playground for a while, and their faces are all red and sweaty. They are so hot! How can the children cool down outdoors?*

| Children's Possible Solutions | Your Role |
| --- | --- |
| 1. One group decides to stop running. The children slow down, then eventually sit down. | Mirror the children's movements as they gradually come to a stop. Help them reflect on the changes in their bodies as they slow down. |
| 2. Other children feel that water might cool them off. They try sipping drinks and using squeeze bottles to squirt water on their faces. | Encourage independence by making materials easily accessible for frequent use. |
| 3. One child says that the sun is too hot. He suggests sitting under the slide or a tree to get away from the sun. | Acknowledge the child's resourcefulness in finding shade. Have the children help you put up a beach umbrella over the sandbox. |

## Extending the Experience

- Create fanning motions with hands, or make folded paper fans.
- Using paintbrushes or watering cans, decorate the playground with cool water art. Check to see what happens when the water dries.
- Make and eat frozen juice pops.

# TRIKE PILEUP

*Children riding their trikes on the hard surface of the playground keep crashing into one another, causing injuries and disagreements about who has the right of way. How can your children make this activity safe and fun?*

| Children's Possible Solutions | Your Role |
| --- | --- |
| **1.** Some children decide to make traffic signs out of cardboard. Using symbols and invented spelling, they create signs like those they have seen, such as one-way signs, stop signs, and yield signs. They put up the signs and try to follow the directions while riding their trikes. | Provide cardboard, markers, and tape. Congratulate the children on their "safe driving." |
| **2.** One child remembers seeing a traffic officer use hand motions to direct cars. The children take turns motioning the trike riders to stop, go one way, and so forth, to avoid crashes. | Help the children clarify their directions by asking open-ended questions such as "How might you motion a rider to make a turn?" Provide a hat or white gloves for the "traffic officer." |
| **3.** Other children decide to make a path for trikes by drawing different-colored chalk arrows and road divider marks on the playground surface. | Offer colored chalk and make wipes available for cleaning hands. |

## Extending the Experience

- Take a neighborhood walk to observe traffic signs.
- Create an obstacle course for trike riders, using materials such as plastic cones or rocks.
- Draw picture maps for the children to follow. Include familiar landmarks, such as buildings or trees.
- Indoors, design a miniature path to negotiate with toy cars, blocks, figurines, or similar objects.

# TERRIFIC TIRES

*A parent drops off a big pile of old tires for your children to play with. What can the children do with them?*

| Children's Possible Solutions | Your Role |
| --- | --- |
| **1.** Several children decide to stack the tires in towers of twos and threes to create a castle to climb in and out of. | Extend dramatic play by asking open-ended questions such as "Where will the princess live?" Compliment the children on their collaboration and physical coordination. |
| **2.** Other children say that it would be fun to see how far the tires will roll. | Challenge the children to try rolling the tires on various surfaces, such as grass or pea gravel, and at different angles, such as on a hill or on a flat trike path. Before each trial, encourage the children to predict what might happen. |

## Extending the Experience

- Place a few tires flat on the ground and fill the centers with sand to make mini-sandboxes.
- Use tires to make an obstacle course. Arrange them for jumping into, hopping around, crawling through, and so forth.
- Roll several tires through puddles. Try to match the tires to the track patterns.
- Make a planter by placing a tire flat on the ground and decorating it with colored tape or paint. Fill the center with dirt and plant seeds or flowers.

# KEEPING DRY

*It has just stopped raining and your children want to go out and play. Everything on the playground is wet. What can the children do to stay dry?*

| Children's Possible Solutions | Your Role |
| --- | --- |
| 1. Several children suggest using sponges and paper towels to wipe off the slide and the swings, just like they do the snack table inside. | Make materials for wiping readily accessible. Compliment the children on applying an indoor idea outdoors. |
| 2. One girl demonstrates how she rolls up the legs of her pants to keep dry. | Comment on the girl's inventiveness. Invite other children to try this idea. |
| 3. Your children discover to their surprise that the equipment inside the storage shed is dry. They decide to play with that equipment rather than use the wet equipment on the playground. | Ask open-ended questions such as "Why do you think these toys are dry?" Encourage the children to think of other materials they like to play with that might also be dry, such as the sand in a covered sandbox. |

## Extending the Experience

- Use colored chalk to draw pictures on construction paper. Blot the pictures on wet grass or dip them in puddles and observe what happens.
- Put on raincoats, and use plastic spray bottles to squirt one another with water.
- Wash doll clothes and hang them up to dry on a clothesline outside.

# CREATIVE CLEANUP

*When outdoor play is over, your children often forget to pick up the sand and water toys, balls, and other playthings. What can the children do to make this chore more enjoyable?*

| Children's Possible Solutions | Your Role |
|---|---|
| 1. A few of the children think that it would be fun to pick up the toys in the wagon and then pull them as fast as they can to the storage shed. | Encourage the children to try out their clever new cleanup idea. |
| 2. Other children want to use different-colored boxes for storing different kinds of toys, just like in the manipulatives area indoors. | Help the children create color-coded boxes: red for balls, blue for water toys, yellow for sand toys, and so forth. Make room for the boxes on low shelves in the storage shed to foster independent cleanup. |
| 3. One child suggests a color game. The leader would call out directions such as "If you're wearing red, pick up the balls." | Provide sufficient cleanup time to put the game into practice. |

## Extending the Experience

- Each week, have Playground Cleanup Day. Together, pick up litter, sticks, and so forth to build pride in how the playground looks.
- Wash dirty toys or trikes with sponges and water to keep them in good shape.
- Beautify the storage shed with painted designs.

# Teacher Resources

**ART SERIES**
*Ideas for successful art experiences.*
Cooperative Art
Special Day Art
Outdoor Art

**BEST OF TOTLINE® SERIES**
*Totline's best ideas.*
Best of Totline Newsletter
Best of Totline Bear Hugs
Best of Totline Parent Flyers

**BUSY BEES SERIES**
*Seasonal ideas for twos and threes.*
Fall • Winter • Spring • Summer

**CELEBRATIONS SERIES**
*Early learning through celebrations.*
Small World Celebrations
Special Day Celebrations
Great Big Holiday Celebrations
Celebrating Likes and Differences

**CIRCLE TIME SERIES**
*Put the spotlight on circle time!*
Introducing Concepts at Circle Time
Music and Dramatics at Circle Time
Storytime Ideas for Circle Time

**EMPOWERING KIDS SERIES**
*Positive solutions to behavior issues.*
Can-Do Kids
Problem-Solving Kids

**EXPLORING SERIES**
*Versatile, hands-on learning.*
Exploring Sand • Exploring Water

**FOUR SEASONS**
*Active learning through the year.*
Art • Math • Movement • Science

**JUST RIGHT PATTERNS**
*8-page, reproducible pattern folders.*
Valentine's Day • St. Patrick's Day • Easter • Halloween • Thanksgiving • Hanukkah • Christmas • Kwanzaa • Spring • Summer • Autumn • Winter • Air Transportation • Land Transportation • Service Vehicles • Water Transportation • Train • Desert Life • Farm Life • Forest Life • Ocean Life • Wetland Life • Zoo Life • Prehistoric Life

**KINDERSTATION SERIES**
*Learning centers for kindergarten.*
Calculation Station
Communication Station
Creation Station
Investigation Station

**1•2•3 SERIES**
*Open-ended learning.*
Art • Blocks • Games • Colors • Puppets • Reading & Writing • Math • Science • Shapes

**1001 SERIES**
*Super reference books.*
1001 Teaching Props
1001 Teaching Tips
1001 Rhymes & Fingerplays

**PIGGYBACK® SONG BOOKS**
*New lyrics sung to favorite tunes!*
Piggyback Songs
More Piggyback Songs
Piggyback Songs for Infants and Toddlers
Holiday Piggyback Songs
Animal Piggyback Songs
Piggyback Songs for School
Piggyback Songs to Sign
Spanish Piggyback Songs
More Piggyback Songs for School

**PROJECT BOOK SERIES**
*Reproducible, cross-curricular project books and project ideas.*
Start With Art
Start With Science

**REPRODUCIBLE RHYMES**
*Make-and-take-home books for emergent readers.*
Alphabet Rhymes • Object Rhymes

**SNACKS SERIES**
*Nutrition combines with learning.*
Super Snacks • Healthy Snacks • Teaching Snacks • Multicultural Snacks

**TERRIFIC TIPS**
*Handy resources with valuable ideas.*
Terrific Tips for Directors
Terrific Tips for Toddler Teachers
Terrific Tips for Preschool Teachers

**THEME-A-SAURUS® SERIES**
*Classroom-tested, instant themes.*
Theme-A-Saurus
Theme-A-Saurus II
Toddler Theme-A-Saurus
Alphabet Theme-A-Saurus
Nursery Rhyme Theme-A-Saurus
Storytime Theme-A-Saurus
Multisensory Theme-A-Saurus
Transportation Theme-A-Saurus
Field Trip Theme-A-Saurus

**TODDLER RESOURCES**
*Great for working with 18 mos–3 yrs.*
Playtime Props for Toddlers
Toddler Art

# Parent Resources

**A YEAR OF FUN SERIES**
*Age-specific books for parenting.*
Just for Babies • Just for Ones • Just for Twos • Just for Threes • Just for Fours • Just for Fives

**LEARN WITH PIGGYBACK® SONGS**
*Captivating music with age-appropriate themes.*
Songs & Games for…
Babies • Toddlers • Threes • Fours
Sing a Song of…
Letters • Animals • Colors • Holidays • Me • Nature • Numbers

**LEARN WITH STICKERS**
*Beginning workbook and first reader with 100-plus stickers.*
Balloons • Birds • Bows • Bugs • Butterflies • Buttons • Eggs • Flags • Flowers • Hearts • Leaves • Mittens

**MY FIRST COLORING BOOK**
*White illustrations on black backgrounds—perfect for toddlers!*
All About Colors
All About Numbers
Under the Sea
Over and Under
Party Animals
Tops and Bottoms

**PLAY AND LEARN**
*Activities for learning through play.*
Blocks • Instruments • Kitchen Gadgets • Paper • Puppets • Puzzles

**RAINY DAY FUN**
*This activity book for parent-child fun keeps minds active on rainy days!*

**RHYME & REASON STICKER WORKBOOKS**
*Sticker fun to boost language development and thinking skills.*
Up in Space
All About Weather
At the Zoo
On the Farm
Things That Go
Under the Sea

**SEEDS FOR SUCCESS**
*Ideas to help children develop essential life skills for future success.*
Growing Creative Kids
Growing Happy Kids
Growing Responsible Kids
Growing Thinking Kids

**THEME CALENDARS**
*Activities for every day.*
Toddler Theme Calendar
Preschool Theme Calendar
Kindergarten Theme Calendar

**TIME TO LEARN**
*Ideas for hands-on learning.*
Colors • Letters • Measuring • Numbers • Science • Shapes • Matching and Sorting • New Words • Cutting and Pasting • Drawing and Writing • Listening • Taking Care of Myself

# Posters

Celebrating Childhood Posters
Reminder Posters

# Puppet Pals

*Instant puppets!*
Children's Favorites • The Three Bears • Nursery Rhymes • Old MacDonald • More Nursery Rhymes • Three Little Pigs • Three Billy Goats Gruff • Little Red Riding Hood

# Manipulatives

**CIRCLE PUZZLES**
African Adventure Puzzle

**LITTLE BUILDER STACKING CARDS**
Castle • The Three Little Pigs

# Tot-Mobiles

*Each set includes four punch-out, easy-to-assemble mobiles.*
Animals & Toys
Beginning Concepts
Four Seasons

**Start right, start bright!**

**Totline products are available at parent and teacher stores.**